José Juan Mateo
Sergi Maicas

Non-Saccharomyces yeasts in wine production

José Juan Mateo
Sergi Maicas

Non-Saccharomyces yeasts in wine production

Isolation, characterization and selection of Hanseniaspora strains

LAP LAMBERT Academic Publishing

Imprint

Any brand names and product names mentioned in this book are subject to trademark, brand or patent protection and are trademarks or registered trademarks of their respective holders. The use of brand names, product names, common names, trade names, product descriptions etc. even without a particular marking in this work is in no way to be construed to mean that such names may be regarded as unrestricted in respect of trademark and brand protection legislation and could thus be used by anyone.

Cover image: www.ingimage.com

Publisher:
LAP LAMBERT Academic Publishing
is a trademark of
Dodo Books Indian Ocean Ltd. and OmniScriptum S.R.L publishing group

120 High Road, East Finchley, London, N2 9ED, United Kingdom
Str. Armeneasca 28/1, office 1, Chisinau MD-2012, Republic of Moldova, Europe
Managing Directors: Ieva Konstantinova, Victoria Ursu
info@omniscriptum.com

Printed at: see last page
ISBN: 978-3-659-49081-1

Non-*Saccharomyces* yeasts contribute to wine production: isolation, characterization and selection of *Hanseniaspora* strains

José Juan Mateo and Sergi Maicas

Departament de Microbiologia i Ecologia

Universitat de València

Acknowledgements

Part of this research has been performed within the Program VLC/Campus, Microcluster IViSoCa. This work was supported by grants from INV-AE112-66049/UV, UV-AE-2006-0239 and INIA RM2007-00001. We wish to thank S. López, C. López, T. Madrigal, A. Romero, C. Thießen and J. Lilao for experimental assistance.

Table of contents

Introduction   5

Chapter I. Yeast classification and identification   7

Chapter II. Contribution to wine by non-*Saccharomyces* yeast. Enzyme producers   15

Chapter III. *Hanseniaspora*   23

Chapter IV. Methodology   25

Chapter V. Results   32

Chapter VI. Discussion   46

Conclusions   51

Bibliography   52

List of figures

Figure 1. Effect of glucose concentration on glycolytic activities of *Hanseniaspora* strains

Figure 2. Effect of ethanol concentration on glycolytic activities of *Hanseniaspora* strains

Figure 3. Effect of temperature on glycolytic activities of *Hanseniaspora* strains

Figure 4. Effect of pH on glycolytic activities of *Hanseniaspora* strains

Figure 5. Effect of glucose, fructose and ethanol concentration on proteolytic activity of *Hanseniaspora* strains

Figure 6. Effect of temperature and pH on proteolytic activity of *Hanseniaspora* strains.

List of tables

Table 1. Teleomorphs, anamorphs and synonyms of some of the non-*Saccharomyces* yeasts in the *Ascomycetous* genera reported on grapes and in wine fermentations

Table 2. *Hanseniaspora* yeast isolates from Utiel-Requena region

Table 3. Results of the RapID Yeast Plus assays

Table 4. Enzymatic activity of the *Hanseniaspora* strains

Table 5. Glycolytic activities of the *Hanseniaspora* strains

Table 6. Effect of buffer and substrate on protease activity

Table 7. Terpene and other volatile compounds in Muscat wine

Introduction

Grape musts naturally contain a mixture of yeast species and wine fermentation is not a "single-species" fermentation. The dominance of *S. cerevisiae* (inoculated or indigenous) in the fermentation is expected and desired. However, the indigenous non-*Saccharomyces* yeasts, already present in the must, and often in greater numbers than *S. cerevisiae*, are adapted to the specific environment and in an active growth state, which gives them a competitive edge (Cray *et al.*, 2013).

It is well established that wine fermentations, as conducted by traditional methods (without inoculation), are not the result of the action of a single species or a single strain of yeast. Rather, the final products result from the combined actions of several yeast species which grow in succession throughout the fermentation process. Previous studies performed in various countries have described the isolation and identification of yeasts from grape surfaces, and quantitative data on the ecology of grape yeasts have concluded that the isolation process of the total yeast population from the grapes is complex and dependent on many factors (Gil *et al.*, 1996; Mendes Ferreira *et al.*, 2001). Fermentations are initiated by the growth of various species of *Candida*, *Debaryomyces*, *Hanseniaspora*, *Hansenula*, *Kloeckera*, *Metschnikowia*, *Pichia* and *Torulaspora*. Their growth is generally limited to the first two or three days of fermentation, after which they die off. Subsequently, the most strongly fermenting and more ethanol tolerant species of *Saccharomyces* take over the fermentation (Pretorius, 2003).

Non-*Saccharomyces* yeasts, as the name suggests, refers to all yeast species found in wine production barring *S. cerevisiae*, with the proviso that this only includes yeast with a positive role in wine production. Recognized spoilage yeasts, such as *Dekkera/Brettanomyces*, are normally left out of this description. Although most fields of research are often focussed primarily on *S. cerevisiae*, non-*Saccharomyces* research can benefit from the techniques and knowledge developed by the *S. cerevisiae* and other yeast researchers (Cray *et al.*, 2013). *S. cerevisiae* yeasts are able to convert sugar into ethanol and CO_2 via fermentation. They have been used for thousands of years by mankind for the production of fermented beverages and foods, including wine. This yeast is adapted to the harsh conditions in grape musts and grapes (high sugar concentration, increasing alcohol concentration, acidity, presence of sulfites, anaerobiosis, and progressive depletion of essential nutrients, such as nitrogen, vitamins, and lipids). But *S. cerevisiae* is not only responsible for the metabolism of grape sugar to alcohol and CO_2 but has an equally important role to play in the formation of secondary metabolites, as well as in conversion of grape aroma precursors to varietal wine aromas (Pretorius *et al.*, 1999; Ribéreau-Gayon *et al.*, 2000; Pretorius, 2003; Swiegers and Pretorius, 2005; Swiegers *et al.*, 2005).

In the past, the influence of non-*Saccharomyces* yeasts in wine was restricted and even eliminated by inoculation with pure *S. cerevisiae* cultures because they have long been regarded as spoilage yeasts (Andorrá *et al.* 2010). However, in the past three decades, great interest has grown in the potential beneficial role of non-*Saccharomyces* yeasts in wine biotechnology (Gil *et al.*, 1996; Mendes Ferreira *et al.*, 2001). It has been shown that some of the metabolites that these yeasts produce may be beneficial and contribute to the complexity of the wine when they are used in mixed fermentations with *S. cerevisiae* cultures (Mateo *et al.*, 1991; Rodriguez *et al.*, 2010b). It is believed that when pure non-*Saccharomyces* yeasts are cultivated with *S. cerevisiae* strains, their negative metabolic activities may not be expressed or could be modified by the metabolic activities of the *S. cerevisiae* strains (Ciani & Comitini, 2011). Several strains belonging to different non-*Saccharomyces* species have been extensively studied in relation to the formation of some metabolic compounds affecting the bouquet of the final product. Moreover some of these yeast showed positive oenological properties and their use in the alcoholic fermentations has been suggested to enhance the aroma and flavor profiles. The non-*Saccharomyces* yeasts have the capability to produce and secrete enzymes in the wine, such as β-glucosidases, which release monoterpenes derived from their glycosylated form. These compounds contribute to the higher fruit-like characteristic of final product.

Chapter I. Yeast classification and identification

Yeast classification

Non-*Saccharomyces* yeast is a loose colloquial term used among wine microbiologists and in wineries, which includes many different yeast species. These yeasts are either basidiomycetous or ascomycetous that have vegetative states which predominantly reproduce by fission or budding and which do not form their sexual states within or on a fruiting body (Kurtzman *et al.*, 2011). Current taxonomies recognize 149 yeast genera comprising nearly 1500 species (Kurtzman *et al.*, 2011). Of these, more than 40 species have been isolated from grape must (Jolly *et al.*, 2006; Ciani *et al.*, 2010). Yeasts may be known by two valid names, the teleomorphic name referring to the sexual state producing ascospores (Kurtzman *et al.*, 2011), and the anamorphic name referring to the asexual state that does not form ascospores. Yeast classification can be difficult because some yeasts do not sporulate easily and the ability to form ascospores can be lost during long-term storage (Kurtzman *et al.*, 2011). Delays between isolation and identification can lead to a recently isolated yeast being identified as either teleomorphic or anamorphic if culture-based techniques are being followed. On-going modifications in yeast taxonomy (Kurtzman *et al.*, 2011) also results in confusion for nontaxonomists. Particularly when citing older literature, it is not always clear what yeasts were essentially investigated. Fortunately, DNA-based approaches have basically helped to clarify modern taxonomy. Some of the more usually encountered teleomorphic yeasts and their anamorphic counterparts in wine and must are shown in Table 1. Oenological non-*Saccharomyces* yeasts can be divided into three groups, (Jolly *et al.*, 2014).

1)   yeasts that are largely aerobic, for example, *Pichia* spp., *Debaryomyces* spp., *Rhodotorula* spp., *Candida* spp., and *Cryptococcus albidus*

2)   apiculate yeasts with low fermentative activity, for example, *Hanseniaspora uvarum* (*Kloeckera apiculata*), *Hanseniaspora guilliermondii* (*Kloeckera apis*), *Hanseniaspora occidentalis* (*Kloeckera javanica*)

3)   yeasts with fermentative metabolism, for example, *Kluyveromyces marxianus* (*Candida kefyr*), *Torulaspora delbrueckii* (*Candida colliculosa*), *Metschnikowia pulcherrima* (*Candida pulcherrima*) and *Zygosaccharomyces bailii*

7

Table 1. Teleomorphs, anamorphs and synonyms (Kurtzman *et al.*, 2011) of some of the non-*Saccharomyces* yeasts in the *Ascomycetous* genera reported on grapes and in wine fermentations. Adapted from Jolly *et al.* (2013).

Teleomorphic form	Anamorphic form	Synonyms[1]
Citeromyces matritensis	*Candida globosa*	
Debaryomyces hansenii	*Candida famata*	*Pichia hansenii*
Dekkera bruxellensis	*Brettanomyces bruxellensis*	
Hanseniaspora guilliermondii	*Kloeckera apis*	
Hanseniaspora occidentalis	*Kloeckera javanica*	
Hanseniaspora osmophila	*Kloeckera corticis*	
Hanseniaspora uvarum	*Kloeckera apiculata*	
Hanseniaspora vineae	*Kloeckera africana*	
Lachancea kluyveri	[3]	*Saccharomyces kluyveri*
Lachancea thermotolerans	[3]	*Kluyveromyces thermotolerans;*
		Candida dattlia
Metschnikowia pulcherrima	*Candida pulcherrima*	*Torulopsis pulcherrima*
Meyerozyma guilliermondii	*Candida guilliermondii*	*Pichia guilliermondii*
Milleronzyma farinosa	[3]	*Pichia farinosa*
Pichia fermentans	*Candida lambica*	
Pichia kluyveri	[3]	*Hansenula kluyveri*
Pichia membranifaciens	*Candida valida*	
Pichia occidentalis	*Candida sorbosa*	*Issatchenkia occidentalis*
Pichia terricola	[3]	*Issatchenkia terricola*
Saccharomycodes ludwigii	[3]	
Starmerela bombicola	*Candida bombicola*	*Torulopsis bombicola*
Torulaspora delbrueckii	*Candida colliculosa*	*Saccharomyces rosei*
Wickerhamomyces anomalus	*Candida pelliculosa*	*Pichia anomala; Hansenula anomala*
Zygoascus meyerae	*Candida hellenica*	
Zygosaccharomyces bailii	[3]	*Saccharomyces bailii*
[2]	*Candida zemplinina*	*Candida stellata* in older literature
[2]	*Candida stellata*	*Torulopsis stellate*

[1]Names found in older literature

[2]No teleomorphic form

[3]No anamorphic form

8

Methods to recover and identify grape yeast species

Several reviews on the analytical approaches to study overall yeast ecology have been published elsewhere (Boundy-Mills, 2006; Ciani *et al.*, 2002; Kurtzman *et al.*, 2011). We will try to extend the discussion to bacterial species and to the issues related to grape analysis, taking in consideration the advice of Lachance (2003) to obtain ecologically meaningful conclusions: (i) adequate sample size, (ii) correct identification, (iii) habitat characterization, (iv) substrate sampling, that is far more important than the sampling procedure, and (v) sample replication, which is more important than serial dilutions and plate counts.

Sampling schemes

Yeast populations in nature suffer fluctuations (Fonseca & Inácio, 2006) that must be taken into account when devising sampling schemes. Ecological conclusions must be based on extensive sampling. In fact, different bunches of grapes bear different populations and it is more informative to analyze several smaller samples than to blend several bunches into a larger one (Barata *et al.*, 2008b). Samples should be taken in several locations in the vineyard so that spatial fluctuations dictated by the uneven microbial distribution are minimized. Repeated sampling over the years is also a sound practice to understand the behavior of natural microbial populations and avoid wrong conclusions.

Sample picking and treatment

Once sampling schemes are defined, the first step in grape analysis concerns the choice of the method to pick the grapes. The main concern should be to aseptically separate sound berries from damaged ones given their completely different microbial load. This cannot be achieved if whole bunches are analyzed because damaged berries may be hidden inside. Further, the number of species of the "sound" bunches would reflect the diversity of the damaged berry and not of the whole sound bunch. Berries may be picked separately in the vineyard, or grape bunches are collected in the field and the berries are separated in the laboratory to give a 300 g sample. When bunches are picked and taken to the laboratory, sound berries should be chosen from fully sound bunches after visual inspection because sound berries from partially damaged bunches bear higher yeast numbers (Barata *et al.*, 2008a).

The isolation from grape juices obtained in the winery (industrial or experimental) may only be regarded as an approximation of the natural grape microbiota. In fact, when grape juice is only sampled in the winery, even at the beginning of fermentation, a different picture of yeasts species may be obtained, due to bulk transport, crushing or pressing in the winery (Barata *et al.*, 2011; Sturm *et al.*, 2006). Therefore, it is mandatory to isolate yeasts from grapes aseptically collected in the vineyard and they must be processed as quickly as possible because grapes damaged during transport to laboratory may accumulate higher numbers of yeasts (Yanagida *et al.*, 1992).

The method for yeast dislodgement from grapes is also of importance. Belin (1972) discussed the efficiency of recovery methods, saying that simple washings with shaking may not be enough. Later, Martini *et al.* (1980) presented evidence of the importance of strong disruptive methods followed by enrichment cultures to obtain an exhaustive picture of the yeast microbiota. Direct enrichment gave lower number of species, than washing and sonication, but enrichment may be the only way to recover fermenting species. These authors advised enrichment in one sample, and agitation or percolation followed by sonication in another sample, but they did not try grape blending. This method was performed by Combina *et al.* (2005) who found higher results with grape blending in plastic bags, than with jet streaming and shaking. Prakitchaiwattana *et al.* (2004) evaluated recovery during 4 successive rinsing for 10 min each. Rinsing released 80% of total yeasts in the first wash and 96% in damaged grapes. Differences in species diversity through all steps were not apparent or were due to a dilution effect. Therefore, the classical food sample suspension in peptone or saline solutions followed by stomaching and serially diluting is a reliable option.

Enrichment cultures

When juice is obtained after grape blending or when suspensions are recovered after grape or single berry washing, the following step is plating or to continue using enrichment cultures. If juice is allowed to ferment spontaneously this corresponds to an auto-enrichment step. True spontaneous fermentations are performed in the absence of sulphur dioxide additions. If this is added, sensitive species are affected. The enrichment may also be done with culture media where single berries, diluted suspensions or juices are introduced.

The enrichment step elicits the recovery of minority species which would not be detected by plating. The typical example is the wine fermenting *S. cerevisiae*, as mentioned before, but the recovery of other technologically relevant species also requires this approach. Renouf & Lonvaud-Funel (2007) only detected the dangerous spoilage yeast *D. bruxellensis* by using a selective medium as an enrichment step. Washing solutions of grapes did not reveal this species. The enrichment step also enabled the detection of *S. cereviriae* and other fermentative species (*C. cantarelli* and *P. fermentans*) while oxidative basidiomycetous species (*Cryptococcus laurentii*, *Rhodotorula mucilaginosa* and *Sporidiobolus pararoseus*) were detected by these authors in the washing solution.

Culture media

The choice of culture media is critical for characterizing complex microbial populations (Loureiro *et al.*, 2004). The utilization of general purpose culture media directly from grape juice or dilutions only enables the recovery of the most frequent and faster growing species. Spread plates using a 0.1 mL sample are commonly preferred and colonies of the most representative species occupy the medium surface and cover minority colonies, leaving undetectable those representing less than about 1% of the population (Fleet *et al.*, 2002).

The recovery of minority species and/or slow growers requires the use of selective media, possibly following sample enrichment. If the enrichment step is done with selective media then a general purpose medium may be used to isolate growing species because the overall populations were already restricted. Wine spoilage species are typically minority species and so it is advised to use selective media. Several reports mention selective media for *Schizosaccharomyces pombe* (Florenzano *et al.*, 1977), *Z. bailii* (Schuller *et al.*, 2000), *D. bruxellensis* (Rodrigues *et al.*, 2001) and *S. cerevisiae* (Kish *et al.*, 1983).

The use of antifungal compounds is advisable to restrict the growth of filamentous fungi, although some yeast species may be affected (Loureiro *et al.*, 2004). Further, the selection of adequate antibiotics is essential to detect separately yeasts, lactic acid and acetic acid bacteria. With appropriate choice of culture media, antibiotics and incubation conditions, it is possible to inhibit, at least partially, the background microbiota and obtain numbers of the selected species.

Incubation conditions and colony selection

After surface inoculation plates should be kept upright when yeasts are to be recovered, as opposed to bacteria (Deák & Beuchat, 1996). Incubation temperature is usually set at 25–30°C, but lower temperatures (10 °C) are essential to recover *S. uvarum* (Sampaio & Gonçalves, 2008). Aerobic conditions are preferred but anaerobiosis was advised for *Sc. pombe* (Florenzano *et al.*, 1977). In liquid media, the availability of oxygen may also be varied by using different shaking rates. Another important issue is to use long incubation periods, up to 14 days, especially when wine spoilage yeasts are screened (Florenzano *et al.*, 1977; Rodrigues *et al.*, 2001; Schuller *et al.*, 2000).

After incubation, strain choice by colony observation must be done by experienced technicians, small differences are frequently overlooked, being also indispensable to perform microscopical examinations. Frequently, same colony morphologies belong to different species, being a good practice to choose more than one colony for each morphological type.

Identification methods

Accurate identification is obviously crucial for the establishment of microbial communities. It is accepted that classical identification techniques based on morphological, biochemical and physiological criteria may have provided incorrect results in the past or could not reach species definition due to heterogeneous phenotypical results. This is particularly true for species of the genus poorly defined by conventional methodologies (e. g. *Candida* spp., *Pichia* spp.), and so it is not surprising that the number of their species is increasing in recent surveys. Moreover, the reproducibility of these techniques is somewhat questionable, since in many cases they depend on the physiological state of the cells. Molecular biological techniques circumvent these difficulties by allowing direct analysis of the genome, irrespective of the physiological state of the cell, providing more precise identifications.

Molecular methods for yeast species identification

The principles underlying yeast identification by molecular techniques have been previously addressed (Giudici & Pulvirenti, 2002). This theme is subjected to a permanent evolution and more recent reviews provide an adequate update of the available methodologies (Cocolin *et al.*, 2011; Fernández-Espinar *et al.*, 2011). The main technical alternatives for species identification are briefly described below, while approaches required for accurate studies of source tracking or evolutionary assessments are available elsewhere (Fernández-Espinar *et al.*, 2011).

Sequencing of ribosomal DNA. Yeast species can be identified by comparison of nucleotide sequences from rDNA regions. The two most commonly used regions are the D1 and D2 regions encoding the 26S (Kurtzman & Robnett, 1998) and 18S (James *et al.*, 1997) ribosomal subunits. The availability of sequences in DNA databases, particularly for the D1/D2 region of the 26S gene, makes this technique useful for assigning unknown yeast to a specific species when the homology of the sequence is greater than 99% (Kurtzman & Robnett, 1998).

Restriction analysis of ribosomal DNA (rDNA). Simpler methods have been designed based on PCR amplification of rDNA regions followed by restriction analysis of the amplified products. A very useful rDNA region that can be used to differentiate between species is that containing the 5.8S gene and the adjacent intergenic regions ITS1 and ITS2. This technique was used by Guillamón *et al.* (1998) for the rapid identification of wine yeasts, and was later extended to yeasts associated with foodstuffs and beverages (de Llanos *et al.*, 2004; Fernández-Espinar *et al.*, 2000). The amplified fragments and restriction profiles for these species with *Hae*III, *Hinf*I, *Cfo*I and *Dde*I are available online at http://yeast-id.com/. In the absence of restriction profiles the solution is to sequence the ribosomal DNA as described before. When restriction profiles are coincident, identification may also rely on the utilization of classical biochemical tests (Barata *et al.*, 2008a; 2008b).

Polymerase chain reaction (PCR)-denaturing gradient gel electrophoresis (DGGE). This genetic fingerprinting technique based on PCR amplification and denaturing gradient gel electrophoresis (DGGE) has been introduced into microbial ecology by Muyzer *et al.* (1993). This technique allows DNA fragments of the same length to be separated on the basis of sequence differences. DNA migration is retarded when the DNA strands dissociate at a specific concentration of denaturing agent. A related technique is temperature gradient gel electrophoresis (TGGE), which is based on a linear temperature gradient for separation of DNA molecules

(Fernández-Espinar *et al.*, 2011). The DGGE and TGGE methods have only recently been used, for yeast identification in wine fermentations (Andorrà *et al.*, 2008; Di Maro *et al.*, 2007; Renouf *et al.*, 2007; Urso *et al.*, 2008). These techniques are directly applied to the sample but their low sensitivity is a major drawback to study minority populations.

Contribution to wine by non-*Saccharomyces* yeast

Ethanol is the central product of alcoholic fermentation. Currently, consumer and market demand for wines containing lower ethanol has shaped research to develop and evaluate strategies to generate reduced- or low-ethanol wines (Kutyna *et al.*, 2010). Numerous studies have reported lower ethanol yields when using non-*Saccharomyces* yeast (Di Maio *et al.*, 2012; Sadoudi *et al.*, 2012). Another alternative to lower ethanol concentration in wine is to exploit the oxidative metabolism observed in some non-*Saccharomyces* species (Gonzalez *et al.*, 2013). Nevertheless, only one study has reported the use of aerobic yeast for the production of reduced alcohol wine. Wines containing 3% v/v ethanol were obtained after fermentation of grape must by *Williopsis saturnus* and *Pichia subpelliculosa* under intensive aerobic conditions. These reduced alcohol wines were considered to be of an adequate quality (Erten & Campbell, 2001).

The variety of flavour compounds produced by diverse non-*Saccharomyces* yeasts is known (Swiegers & Pretorius, 2005; Swiegers *et al.*, 2005). The metabolic products generated from non-*Saccharomyces* growth include terpenoids, esters, higher alcohols, glycerol, acetaldehyde, acetic acid and succinic acid (Fleet *et al.*, 1984; Clemente-Jimenez *et al.*, 2004). The primary flavour of wine is derived from the grapes, while secondary flavours are derived from ester formation by yeasts during wine fermentation (Lambrechts & Pretorius, 2000). Several flavour and aroma compounds in grapes are present as glycosylated flavourless precursors (Pretorius, 2003). These compounds may be hydrolysed by the enzyme β-glucosidase to form free volatiles that can increase the flavour and aroma of wine, but this enzyme is not encoded by the *S. cerevisiae* genome (Ubeda-Iranzo *et al.*, 1998). In contrast, non-*Saccharomyces* yeasts belonging to the genera *Debaryomyces*, *Hansenula*, *Candida*, *Pichia* and *Kloeckera* possess various degrees of β–glucosidase activity and can play a role in releasing volatile compounds from non-volatile precursors (Rosi *et al.*, 1994; Spagna *et al.*, 2002). Co-fermentation of Chardonnay grape juice with *Debaryomyces pseudopolymorphus* and *S. cerevisiae* resulted in an increased concentration of the terpenols: citronellol, nerol and geraniol in wine (Cordero Otero *et al.*, 2003). Similarly, cofermentation of Muscat grape juice with *Debaryomyces vanriji* and *S. cerevisiae* produced wines with increased concentration of several terpenols (Garcia *et al.*, 2002). Equally, mixed cultures of Sauvignon Blanc grape juice with *C. zemplinina/ S. cerevisiae* and *T. delbrueckii/S. cerevisiae* produced wines with high concentrations of terpenols compared to wines only fermented with *S. cerevisiae* (Sadoudi *et al.*, 2012).

Another strategy to increase the release of bound volatile compounds is to exogenously add enzyme preparations that can act on nonvolatile precursors. Numerous studies have characterized and described the effect of β-glucosidase addition on grape juice or wine, focusing particularly in the inhibition of β-glucosidase activity by sugar, alcohol, pH and/or temperature. An intracellular β-glucosidase from *Debaryomyces hansenii*, which is not inhibited by glucose and ethanol, was used during fermentation of Muscat grape juice resulting in an increase in concentration of monoterpenols in the wine (Yanai & Sato, 1999). The concentration of volatile terpenes in Arien, Riesling and Muscat wines was also increased following addition of an enzyme extract from *Debaryomyces pseudopolymorphus*. Therefore, sensory differences were found between actions (Arevalo-Villena *et al.*, 2007b). Over 160 esters have been distinguished in wine (Jackson, 2000). These esters can have a helpful effect on wine quality, especially in wine from varieties with neutral flavours that are consumed shortly after manufacture (Lambrechts & Pretorius, 2000). Non-*Saccharomyces* can be divided into two groups, neutral yeasts (producing little or no flavour compounds) and flavour-producing species. Flavour-producing yeasts included *P. anomala* (*Hansenula anomala*) and *K. apiculata*. *Candida pulcherrima* is also known to be a high producer of esters (Clemente-Jimenez *et al.*, 2004). The net accumulation of esters in wine is determined by the balance between the yeast's ester-synthesizing enzymes and esterases (responsible for cleavage and in some cases, formation of ester bonds) (Swiegers & Pretorius, 2005). Although extracellular esterases are known to occur in *S. cerevisiae* (Ubeda-Iranzo *et al.*, 1998), the situation for non- *Saccharomyces* needs further investigation. Different non- *Saccharomyces* yeasts produce different levels of higher alcohols (n-propanol, isobutanol, isoamyl alcohol, active amyl alcohol) (Lambrechts & Pretorius, 2000). This is important during wine production, as high concentrations of higher alcohols are generally not desired, whereas lower values can add to wine complexity.

Glycerol, the next major yeast metabolite produced during wine fermentation after ethanol, is important in yeast metabolism for regulating redox potential in the cell (Scanes *et al.*, 1998). Glycerol contributes to smoothness (mouth-feel), sweetness and complexity in wines, but the grape variety and wine style will govern the extent to which glycerol impacts on these properties (Ciani & Maccarelli, 1998). Although the quality of Chardonnay, Sauvignon Blanc and Chenin Blanc is not enhanced by increased glycerol concentrations (Nieuwoudt *et al.*, 2002), some wines might benefit from increased glycerol levels. Several non-*Saccharomyces* yeasts, particularly *L. thermotolerans* and *C. zemplinina*, can consistently produce high glycerol concentrations during wine fermentation (Comitini *et al.*, 2011). Unfortunately, increased glycerol production is usually linked to increased acetic acid production (Prior *et al.*, 2000), which can be detrimental to wine quality. Spontaneously fermented wines have higher glycerol levels, indicating a possible contribution by non- *Saccharomyces* yeasts (Henick-Kling *et al.*, 1998).

However, the use of some non-*Saccharomyces* yeast in mixed fermentations with *S. cerevisiae* can generate wines with decreased volatile acidity and acetic acid concentration (Comitini *et al.*, 2011). Some non-*Saccharomyces* yeasts are able to form succinic acid (Ciani & Maccarelli, 1998). This correlates with high ethanol production and ethanol tolerance. Succinic acid production could positively influence the analytical profile of wines by contributing to the total acidity in wines with insufficient acidity. Nevertheless, succinic acid has a 'salt-bitter-acid' taste and excessive levels will negatively influence wine quality. Other non-*Saccharomyces* metabolites can act as intermediaries in aroma metabolic pathways. Acetoin is considered a relatively odourless compound in wine (Romano & Suzzi, 1996). However, diacetyl and 2,3-butanediol (potentially off-flavours in wine) can be derived from acetoin by chemical oxidation and yeast-mediated reduction, respectively. This indicates that acetoin can play a role in off-flavour formation in wines. Definitely, high concentrations of acetoin produced by non-*Saccharomyces* yeasts can be utilized by *S. cerevisiae* in mixed and sequential culture fermentations (Zironi *et al.*, 1993).

Other compounds that are known to play a role in the sensory quality of wine include volatile fatty acids, carbonyl and sulphur compounds (Lambrechts & Pretorius, 2000). There are over 680 documented compounds in wine and a large number of these can, depending on concentration, contribute either positively or negatively to wine aroma and flavour. Volatile thiols greatly contribute to the varietal character of some grape varieties, particularly Sauvignon Blanc (Swiegers *et al.*, 2009). Some non-*Saccharomyces* strains, specifically isolates from *C. zemplinina* and *Pichia kluyveri* can produce significant amounts of the volatile thiols 3-

mercaptohexan- 1-ol (3MH) and 3-mercaptohexan-1-ol acetate (3MHA), respectively, in Sauvignon Blanc wines (Anfang *et al.*, 2009). Similarly, *T. delbrueckii*, *M. pulcherrima* and *L. thermotolerans* have also been described as able to release important quantities of 3MH from its precursor during Sauvignon Blanc fermentation (Zott *et al.*, 2011). Other non-*Saccharomyces* extracellular enzymatic activities, such as proteolytic and pectinolytic (polygalacturonase) enzymes, might also be beneficial to winemaking (Strauss *et al.*, 2001). For example, proteolytic activity of some non-*Saccharomyces* yeast could lead to a reduction in protein levels with accompanying increase in protein stability of the end-product. Species found to produce the greatest number of extracellular enzymes are *C. stellata*, *H. uvarum* and *M. pulcherrima*. Non-*Saccharomyces* yeasts have also been reported to affect the concentration of polysaccharides in wine (Domizio *et al.*, 2011). Polysaccharides can positively influence wine taste and mouth-feel by increasing the perception of wine 'viscosity' and 'fullness' on the palate (Vidal *et al.*, 2004). The early death of some non-*Saccharomyces* yeasts during fermentation can also be a source of specific nutrients for S. cerevisiae enabling it to ferment optimally. These nutrients include cellular constituents such as cell wall polysaccharides (mannoproteins). For this method of nutrient supply to be effective, any killer or other inhibitory effects by the non-*Saccharomyces* yeasts against *S. cerevisiae* should be known (Fleet *et al.*, 2002) so that the subsequent *S. cerevisiae* fermentation is not adversely affected.

Non-*Saccharomyces* strains as enzyme producers for vinification

Glycosidases

Research over the last decades has revealed that a great number of plant tissues flavour compounds are glycosilated and accumulate as non-volatile and flavourless glycoconjugates (Mateo & Jimenez, 2000). Although results in literature had long suggested the occurrence of glycosidically bound flavor compounds in plants, the first clear evidence was found in 1969 by Francis & Allock in rose flowers (Francis & Allock, 1969). The work of Cordonnier & Bayonove (1974) suggesting the occurrence in grapes of monoterpenes, important flavour compounds, as glycoconjugates on the basis of enzymatic works was later confirmed by identification of glycosides (Williams *et al.*, 1982). These findings opened a new field of intensive research on the chemistry of glycoconjugated flavour compounds to exploit this important flavour source present in both plants and fruit tissues. Some aglycones are already odorous when released from glycosides. They can therefore contribute to the floral aroma of some wines (Mateo & Jimenez, 2000), grapes (Gunata *et al.*, 1993), apricots (Chairote *et al.*, 1981), peaches (Engel *et al.*, 1988) and tea (Ogawa *et al.*, 1997). This is the case of monoterpenes such as geraniol, nerol and linalool which possess mainly floral attributes and low odour thresholds (100-400 ppb) (Rapp & Mandery, 1986).

Terpene compounds belong to the secondary plant constituents, of which the biosynthesis begins with acetyl-CoA (Manitto, 1980). Microorganisms are also able to synthesize terpene compounds (Hock *et al.*, 1984) but the formation of terpenes by *Saccharomyces cerevisiae* has not yet been observed (Rapp & Mandery, 1986). Several authors have shown that terpenes play a significant role in the varietal flavour of wines by means of their transformation to other compounds (Wilson *et al.*, 1986).

Terpene glycosides can also be hydrolysed by an enzymatic way, a more interesting way because it produces a more "natural" flavour in the wine (Mateo & Jimenez 2000; Gunata 2002). The glycosidase flavour potential from grape remains unfortunately quite stable during winemaking and in young wines as well. So, to enrich wine flavour by release of free aromatic compounds from natural glycoside precursors, particularly pathways are required. Mainly, enzymatic hydrolysis of glycosides is carried out with various enzymes which act sequentially according to two steps: firstly, α-L-rhamnosidase, α-L-arabinosidase or β-D-apiosidase make the cleavage of the terminal sugar and rhamnose, arabinose or apiose and the corresponding β-D-glucosides are released; subsequently liberation of monoterpenol takes place after action of a β-D-glucosidase (Gunata *et al.*, 1988). Nevertheless, one-step hydrolysis of disaccharide glycosides has also been described; enzymes catalysing this reaction have been isolated from tea leaves (Ogawa *et al.*, 1997) and grapes (Gunata *et al.*, 1998). This one-step reaction occurs through the cleavage of the aglycone linkage which yields a disaccharide and aglycone, the identity of which have been confirmed by HPLC and GC/MS (Gunata *et al.*, 1998).

Enzymatic hydrolysis of glycoside extracts from Muscat, Riesling, Semillon, Chardonnay, Sauvignon and Sirah varieties have provoked the liberation not only of terpenes, but also C-13 norisoprenoids, such as 3-oxo-α-ionol and 3-hydroxy-β-damascenona (Gunata *et al.*, 1990a). These compounds are totally glycosilated in the grape and, opposite with terpenes, they are found in the same quantities in all the grape varieties, aromatics or neutral, and they are capable of awarding certain typicity to the wine flavour because they have lower threshold values than terpenes and they contribute characteristic aromatic features (Razungles *et al.*, 1987).

Yeasts of the *Hansenula* species isolated from fermenting must were reported to have an inducible β-glucosidase activity, but this enzyme was inhibited by glucose (Grossman *et al.*, 1987). Other yeast strains such as *Candida molischiana* (Gonde *et al.*, 1985) and *C. wickerhamii* (Leclerc *et al.*, 1984) also possess activities towards various β-glucosides and they were little influenced by the nature of aglycon (Gunata *et al.*, 1990b). β-Glucosidase from *C. molischiana* was immobilized to Duolite A-568 resin, showing similar physicochemical properties to those of free enzyme. The immobilized enzyme was found to be very stable under wine conditions and could be used repeatedly for several hydrolyses of bound aroma (Gueguen *et al.*, 1997b). *Endomyces fibuliger* also produces extracellular β-glucosidase when grown in malt extract broth (Brimer *et al.*, 1998).

Screening 370 strains belonging to 20 species of yeasts, all of the strains of the species *Debaryomyces castelli, D. hansenii, D. polymorphus, Kloeckera apiculata* and *Hansenula anomala* showed β-glucosidase activity (Rosi *et al.*, 1994). A strain of *D. hansenii* exhibited the highest exocellular activity and some wall-bound and intracellular activity and its synthesis, occurred during exponential growth, was enhanced by aerobic conditions and repressed by high glucose concentration. The optimum condition for this enzyme was pH 4.0-5.0 and 40°C. This enzyme was immobilized using a one-step procedure on hydroxyapatite. The immobilized enzyme exhibited a lower activity than the purified free enzyme, but was much more stable than the enzyme in cell-free supernatant (Riccio *et al.*, 1999). Their studies have shown the ability of several wine yeasts to hydrolyse terpenoids, norisoprenoids and benzenoids glycosides; among wine yeasts *Hanseniaspora uvarum* was able to hydrolyse both glycoconjugated forms of pyranic and furanic oxides of linalool (Fernandez *et al.*, 2003) Other authors have also shown the important role of non-*Saccharomyces* species in releasing glycosidic bound fraction of grape aroma components (Mendes *et al.*, 2001)

Finally, the situation regarding *S. cerevisiae* is more complex because this yeast is capable to modify the terpenic profile of the wine; so, it can produce citronellol from geraniol and nerol, the intensity of this transformation depends on the yeast strain used (Hernandez *et al.*, 2003). Other authors propose a more complex scheme: geraniol was transformed by these yeasts into geranyl acetate, citronellyl acetate and citronellol, while nerol was transformed into neryl acetate;

in addition, geraniol was transformed into linalool and nerol was cyclized to α-terpineol at must pH (Di Stefano *et al.*, 1992).

Few data are available regarding glycosidase activities of oenological yeast strains and the technological properties of the enzymes. Low α-rhamnosidase, α-arabinosidase or β-apiosidase activities were detected in *S. cerevisiae* (Delcroix *et al.*, 1994). Nevertheless, data on β-glucosidase activity on *Saccharomyces* are contradictory. First results showed that these yeasts had a very low activity (Gunata *et al.*, 1990c) but Delcroix *et al.* (1994) found three enological strains showing high β-glucosidase activity. On the other hand, Darriet *et al.* (1988) have shown that oxidases located in the periplasmic space of a strain of *S. cerevisiae* were able to hydrolyse monoterpene glucosides of Muscat grapes; they found also that the activity of this β-glucosidase was glucose independent. Mateo and Di Stefano (1997) detected β-glucosidase activity in different *Saccharomyces* strains on the basis of its hydrolytic activity on *p*-nitrophenyl-β-D-glucoside (*p*NPG) and terpene glucosides of Muscat juice. This enzymic activity is induced by the presence of bound β-glucose as carbon source in the medium and seems to be a characteristic of the yeast strain. This β-glucosidase is associated with the yeast cell wall, is quite glucose independent but is inhibited by ethanol. These results could open new pathways regarding other glycosidase activities in *S. cerevisiae*; α-rhamnosidase, α-arabinosidase or β-apiosidase activities could be induced in wine yeast by changing the composition of the medium including inductive compounds, as well as in filamentous fungi (Dupin *et al.*, 1992).

Proteases

Non-*Saccharomyces* strains are also used in vinification regarding their ability to produce several enzymes (Mateo *et al.*, 2011). Our research has also been focused to the study of these enzymes, particularly proteases. Proteases are categorized on the basis of their catalytic mechanism, the amino acid residues present in the catalytic site and their three-dimensional structure. According to the NC-IUBMB, proteases can be categorized into four mechanistic classes which include the serine endopeptidases, cysteine endopeptidases, aspartic endopeptidases and metalloendopeptidases. Each type of protease has a specific ability to break a certain peptide bond and exhibits a characteristic set of functional amino acid residues arranged in a specific configuration to produce its catalytic site (Barrett *et al.*, 2004; Tyndall *et al.*, 2005). The aspartic proteases secreted by non-*Saccharomyces* yeasts have a tertiary structure consisting of two approximately symmetric lobes with each lobe carrying an aspartic acid residue to form the catalytic site. In contrast to other types of proteases, the activity of the aspartic proteases is dependent on pH conditions (Borrelli *et al.*, 2008).

The ability of yeasts to release proteases has been observed by many researchers because of their potential to degrade haze proteins in wine and to generate nutrient sources for microorganisms (Lagace & Bisson, 1990). Protein haze is one of the most important changes for alcoholic beverages producers. This phenomenon occurs in juice with low polyphenol content as a result of coagulation of proteins in alcoholic beverage from unfavorable storage conditions, resulting in their aggregation. The denatured proteins can either precipitate to form an amorphous sediment or deposit, or can flocculate producing a suspended unstable and unsightly haze in bottle (Pocock & Waters, 2006). The presence of haze reduces the commercial value of the product making it unacceptable for consumers because it may be perceived as microbial spoilage (Waters *et al.*, 2005). Typically in industry, the haze caused by proteins is removed from wine by bentonite fining but, under certain conditions, it may have an adverse effect on the quality of beverage because some colour, flavor and aroma compounds may be removed together with proteins (Waters *et al.*, 2005). Because of the drawbacks presented by this treatment, alternative methods to remove haze-causing proteins have been investigated, amongst these the application of proteolytic enzymes (Rosi & Costamagna, 1987). Dizy & Bisson (2000) demonstrated that strains of *Hanseniaspora* produced the most proteolytic activity in juice and affected the protein profile of the finished product.

Besides the potential to aid in haze reduction, the extracellular proteolytic activity of non-*Saccharomyces* yeasts may also hold potential to increase the assimilable nitrogen sources for the grown of microorganisms during fermentation (Cramer *et al.*, 2002). Insufficient initial assimilable nitrogen sources may lead to stuck or sluggish fermentations (Henschke & Jiranek, 1993). On the other hand, compounds contributing to the fermentation bouquet of beverages, such as esters, higher alcohols and volatile fatty acids arise as primary metabolites of yeast sugar and aminoacid metabolism (Swiegers *et al.*, 2005).

Hanseniaspora, yeasts mainly found in soil, on fruits and trees and in spoiled foods and beverages are characterized by apiculate cells with vegetative reproduction by bipolar budding in basipetal succession. There are several species in the genus *Hanseniaspora* (Cadez *et al.*, 2003; Jindamorakot *et al.*, 2009; Kurtzman *et al.*, 2011). They are physiologically very similar (Kurtzman & Robnett, 1998); they ferment glucose, assimilate few carbon compounds (arbutin, cellobiose, glucose, glucono-δ-lactone and salicin), and require inositol for growth (Kurtzman *et al.*, 2011). The identification of yeast species has been traditionally based on assimilation and fermentation tests and morphological traits. These studies must be accompanied by other assays to assess a correct and unambiguous identification of the yeasts. Molecular methods have been used to explore this yeast biodiversity, which confer a high degree of accuracy in the final identification (Kurtzman *et al.*, 1994; 2011; Guillamón *et al.*, 1998; Esteve-Zarzoso *et al.*, 1999; Arroyo-López *et al.*, 2006).

From a biotechnological point of view, although these yeasts can produce spoilage of fruits (Arroyo-López *et al.*, 2008), they also possess many interesting technological properties (Charoenchai *et al.*, 1997, Strauss *et al.*, 2001; Maicas & Mateo, 2005). In this way, exocellular enzymes of relative yeast species in different ecosystems have been previously reviewed (Mateo *et al.*, 2011). Winemaking has been the process more deeply studied regarding the influence of *Hanseniaspora* (*H. guilliermondii, H. uvarum, H. osmophila* and *H. vineae*), and other non-*Saccharomyces* species. The modification of the characteristics of the wine is attributed to the capacity of certain non-*Saccharomyces* yeasts to produce and secrete hydrolytic enzymes able to transform grape compounds (Pando *et al.*, 2012). These compounds are present in varying amounts as non-volatile flavor glycosylated precursors (Ugliano, 2009), mainly disaccharides 6-O-α-L-arabinofuranosyl-β-D-glucopyranoside, 6-O-α-L-rhamnopyranosyl-β-D-glucopyranoside, 2-O-β-D-xylosyl-D-glucopyranoside and 6-O-β-L-apiofuranosil-β-D-glucopyranoside (Maicas & Mateo, 2005). The action of enzymes produced by wine yeasts i.e. β-glucosidase or β-xylosidases can contribute to liberate flavor from these compounds (Romano *et al.*, 2003). Several groups have already made trial fermentations to study the compounds generated by *Hanseniaspora* (Paraggio, 2004; Arévalo *et al.*, 2007a; Swangkeaw, 2011).

In the present study the selection and correct identification of *Hanseniaspora* strains to be further used as a source of enzymes in winemaking industry and also in other biotechnological processes is described. These yeasts are non-pathogenic and are recognized as safe organisms (GRAS) that can be used for production of a variety of enzymes. Finally, our essays have proved the benefits for using these yeasts to increase the level of terpenes in Muscat wine.

Chapter IV. Methodology

Yeast strains

Grape juice and wine samples were collected from cellars from the D.O. Utiel-Requena (Eastern Spain) during the last 22 years (Gil *et al.*, 1996; Mateo *et al.*, 1991; 1992; 2011). Samples were stored at -20°C and routinely spread onto Malt Agar (20 g/L malt extract, 1 g/L peptone, 20 g/L glucose, 20 g/L agar) and grown at 28°C for viability check, or in YPD (20 g/L yeast extract, 10 g/L peptone, 20 g/L glucose) for routinely assays. A total of 31 strains of *Hanseniaspora* yeasts strains from our collection, previously selected by using lysine agar as differential medium (Heard & Fleet, 1986) were used in this study (Table 2).

Table 2. *Hanseniaspora* yeast isolates from Utiel-Requena region

Yeast species	Number of isolates	Matching nucleotides (%)[a]	Collection number
Hanseniaspora uvarum	18	99.4-99.6%	E18, E20, E44,F42, G09, G89, H03, H105, H106, H107, H156, H172, B21,B95, M45, M46, M47, M48
H. vineae	4	99.5-99.6%	E71,G26, P30, P38
H. guilliermondii	8	99.6-99.8%	A27,A29, B02, B04, C06,D55, D56, D57
H. osmophila	1	99.7%	C07
Total	31		

[a]Sequence identity in the D1/D2 region of 26S ribosomal gene and closest relative species in the NCBI GenBank database.

Isolates were identified according to their morphological and physiological characteristics described by Kurtzman *et al.* (2011) at a first step, and therefore following molecular techniques.

Yeast typing by physiological techniques

The yeast identification system Rapid ID Yeast Plus System (Remel, Lenexa, USA) was used to assess the strain-specific pattern of carbon compound assimilation. Strips were inoculated and inspected for growth daily, up to 7 days, according to the manufacturer's instructions.

Yeast typing by molecular techniques

DNA extraction

Pure cultures of yeasts were grown on 10 mL of YPD medium at 28 °C for 24-48 h on an orbital shaker. Small-scale preparation of chromosomal DNA was performed by using Ultraclean Microbial DNA isolation Kit (MoBio, Carlsbad, CA). The quality of the extracted DNA was checked by electrophoresis in 0.8 % (w/v) agarose mini-gels using TBE (45 mM Tris borate, 1 mM EDTA, pH 8.0) buffer with added ethidium bromide at a final concentration of 0.5 μg/mL. These gels were used to estimate the approximate DNA concentration. An amount of 10% of the preparation was compared to digested lambda DNA cleaved with *Eco*RI and *Hin*dIII (Boehringer Mannheim GmbH, Mannheim, Germany).

PCR amplification

The region between the genes 18S rRNA and 28S rRNA was amplified using Internal Transcribed Spacers ITS1 (5'-TCCGTAGGTGAACCTGCGG-3') and ITS4 (5'-TCCTCCGCTTATTGATATGC-3') as primers (Boehringer Mannheim GmbH, Mannheim, Germany) (White *et al.*, 1990). The amplification reaction was performed in a Primus 25 thermocycler (MWG, Ebersberg, Germany) under the following conditions: a 50-μL reaction mixture was prepared with 1.5 U *Taq* DNA polymerase (Boehringer Mannheim GmbH, Mannheim, Germany), 0.5 μM of each primer, 0.2 mM of each dNTP, 10 mM Tris-HCl, 1.5 mM MgCl$_2$, 50 mM KCl (buffer), and 10-20 ng yeast DNA. The mixture was subjected to an initial denaturing step of 5 min at 95 °C, followed by 35 cycles of 1 min at 95 °C, 2 min at 52 °C, and 2 min at 72 °C, and a final extension step of 10 min at 72 °C.

Amplification products were separated by electrophoresis in a 0.8% (w/v) agarose gel. A 25bp DNA ladder commercial standard (Takara, Shiga, Japan) was used. PCR amplifications were purified and washed with a High-pure PCR product amplification kit (Boehringer Mannheim GmbH, Mannheim, Germany).

Restriction analysis

PCR products were digested without subsequent purification with the following restriction enzymes: *Cfo*I, *Hae*III, *Hinf*I and *Msp*I for differentiation of yeasts to species or strain level. The final digestion volume was 20 μL, containing 12 μL of amplified DNA solution, 2 μL of restriction enzyme (10 U/μl) (Boehringer Mannheim GmbH, Mannheim, Germany), 2 μL of restriction buffer supplied by Boehringer Mannheim GmbH, and 4 μL of purified water. The mixture was incubated at 37 °C for 8 h. The restriction fragments were checked by electrophoresis in 1.2% (w/v) agarose gel. The restriction fragment sizes measured as base pairs were calculated in comparison with a 100 bp DNA ladder commercial standard (Takara, Shiga, Japan). Profiles were compared with the Yeast-id.org database (CECT, Paterna, Spain).

Sequence analysis of the D1/D2 domains of the 26S rDNA gene

This procedure was used to confirm the identifications previously carried out by RFLP analysis of the 5.8S-ITS rDNA region. PCR amplification for the D1/D2 domain of 26S rDNA was basically performed according to Kurtzman and Robnett (1998). The PCR product was purified using UltraClean PCR Clean Up kit (MoBio, CA, USA) according to the manufacturer's instructions. Direct sequencing of the purified PCR products was performed by ABI Prism BigDye Terminator Cycle Sequence Ready Reaction Kit (Applied Biosystems, Stafford, TX). The sequences were aligned, by using the BLAST program, with complete or nearly complete 26S rDNA gene sequences retrieved from the EMBL nucleotide sequence data libraries.

Qualitative screening of the biochemical activities

β-Glucosidase activity

The basal medium consisted on 1.7 g/L Yeast Nitrogen Base (Difco), 5 g/L ammonium sulphate, 5 g/L glucose and 20 g/L agar. After autoclaving, 2 mL of a sterile 1% (w/v) 4-methylumbelliferyl-β-D-glucopyranoside (Sigma, St. Louis, MO) were added to 100 mL of melted medium (Manzanares *et al.,* 1999). The presence of the enzymatic activity was visualized as a fluorescent halo surrounding yeast growth by plate exposition to UV light.

Xylanase activity

Xylanase production was determined according to Hernández *et al.* (2007) by spreading yeast colonies onto agar plates containing 5 g/L xylan, 5 g/L peptone and 5 g/L NaCl. Plates were incubated at 28°C for 7 days. A clear zone around the colony was indicative of xylanase activity.

Pectinase and polygalacturonase activities

The assays were carried out in the following medium: 1 g/L yeast extract, 1 g/L ammonium sulphate, 6 g/L $NaHPO_4$, 3 g/L KH_2PO_4, 5 g/L pectin (for pectinase activity) or 5 g/L polygalacturonic acid (for polygalacturonase activity), 15 g/L agar. After streaking 48 h-old yeast cultures onto the surface of the medium, the plates were incubated at 28°C for five days and then revealed by addition of a solution of hexadecyltrimethylammonium bromide (1 g/L) as described by Oliveira *et al.* (2009). Both activities were evidenced by the presence of a clear halo around the colonies.

Lipase and esterase activities

Yeast isolates were used to determine esterase and lipase activities on tributyrin and rhodamine olive-oil agar media, respectively, according to the previously described procedures (Rodríguez *et al.*, 2010; Madrigal *et al.,* 2013). After 48 h of incubation at 28°C in the media, colonies were investigated. For detection of lipase activity, they were irradiated with UV light at 350 nm; lipase activity was detected by a fluorescent halo around colonies. Esterase activity was detected by the formation of a halo around colonies. *Saccharomyces cerevisiae* CECT11783 and *Candida molischiana* ATCC2516 were used as a positive control for polygalacturonase and β-glucosidase production, respectively. Lipase from *C. antarctica,* esterase from *S. cerevisiae* and pectinase from *Aspergillus niger* from Sigma, St. Louis, MO were used for the same purpose.

Protease activity

Exocellular protease production was determined by spreading yeast colonies onto YPD agar plates containing 20 g/L casein. Plates were incubated at 28°C for 7 days. A clear zone around the colony is indicative of protease activity.

Quantitative spectrophotometric assays

β-Glucosidase and β-xylosidase activities

β-glucosidase and β-xylosidase activities were basically assayed using 4-nitrophenyl-β-D-glucopyranoside and 4-nitrophenyl-β-D-xylopyranoside as the substrates (Romero *et al.*, 2012). The yeasts were centrifugated and resuspended in 750 μL of 0.2 M citrate-0.1 M phosphate buffer (pH 5.0). Then 250 μL of 5 mM pNP-substrate in the same buffer was added and the mixture was incubated at 40°C for 90 min. The reaction was stopped by adding 1.0 mL of 0.2 M Na_2CO_3, and absorbance at 404 nm was measured. Activity was expressed as nanokatals (1 nkat = 1 nmol of pNP liberated in 1 hour by 10^6 yeasts).

Quantitative protease activity

The protease assay procedure was carried out as described by Tremacoldi and Carmona (2005) with modifications. Briefly, yeasts were inoculated in YPD-casein medium and incubated at 26°C for 6 days to induce protease activity. After adding 10^6 cells/mL to a solution containing 500 μL 0.1 M citrate phosphate buffer pH 6.0 and 500 μL 1 % (w/v) BSA in the same buffer, mixture was incubated 1 h at 50°C and then, 1 mL 3 % (w/v) TCA was added. Tubes were maintained on ice for 30 min and centrifugated. Liberated tyrosine was determined by absorbance at 280 nm. The protease activity was calculated as nanokatals (1 nkat = 1 nmol of tyrosine liberated in 1 hour by 10^6 yeasts).

Effect of sugars and ethanol on β-glucosidase and β-xylosidase activities

Effects of sugar on enzymes activities were conducted by using sugar concentrations over a range of 0-500 mM (glucose and fructose) or 0-50 mM (sucrose). Effect of ethanol (Merck,

Darmstadt, Germany) on enzyme activity was conducted by using ethanol concentration over a range of 0-20% (v/v). The enzyme assays were carried out as previously described.

Effect of temperature and pH on β-glucosidase and β-xylosidase activities

The pH optimum was determined in 0.2 M citrate-0.1 M phosphate buffer covering a pH range from 3.0 to 8.0, at 40°C for 90 min. The temperature optimum was measured from 4°C to 60°C for 90 min of incubation in the same buffer, at pH 5.0. The residual activities were determined.

Effect of sugars and ethanol on protease activity

Effects of different sugars (glucose, fructose) on enzyme activities were conducted by using sugar concentrations over a range of 0-500 mM. Effect of ethanol (Merck, Darmstadt, Germany) on enzyme activity was conducted by using ethanol concentration over a range of 0-20% (v/v). The enzyme assays were carried out as previously described.

Effect of temperature and pH on protease activity

The pH optimum was assayed in 0.2 M citrate-0.1 M phosphate buffer covering a pH range from 3 to 8, at 40°C for 90 min. The temperature optimum was measured from 4°C to 60°C for 90 min of incubation in the same buffer (pH 5.0). The residual activities were determined.

Winemaking

Muscat juice (1.000 hL) was fermented using 30 g/hL commercial wine yeast strain *S. cerevisiae* QA23 (Lalvin, Lallemand, Montreal). Fermentation was carried out at 16-24 °C. Must sampling for analysis of sugar concentrations were performed weekly. After 23 days, when less than 2 g/L residual sugar remained, the wine was separated from the gross lees. Samples were collected and stored at 4 °C for the second inoculation with *Hanseniaspora*. This wine (from Turís winery) was as follows: ethanol, 13.3% v/v; pH, 3.33; titratable acidity, 3.6 g/L; volatile acidity, 0.35 g/L; malic acid, 1.0 g/L. Wine was sterilized by 0.45-μm filtration and sterility was verified by 100-μL wine spreading on YPD plates.

Determination of volatile compounds liberated from wine incubated with *Hanseniaspora* strains

H. uvarum H107 and *H. vineae* G26 and P38 were grown in YPD, centrifugated, resuspended in distilled water, added to 400-mL Muscat wine at a final concentration of 1.0×10^6 cfu/mL and incubated in separated 500-mL cotton-plugged flasks at 20°C for 14 days, without shaking. Wine produced with *S. cerevisiae*, without added *Hanseniaspora* yeasts, was assayed in parallel to act as a control. Then, the isolation of terpenes was carried on C18 SPE columns (Waters, Milford, MA). 250 mL of each wine were eluted through the columns, previously activated with 30 mL of methanol (Merck, Darmstadt, Germany) followed by 50 mL of water. The cartridge was washed with 100 mL of water and lipophilic compounds (terpenes) were eluted with 50 mL of dichloromethane (Merck). Volatile compounds were determined by using an Agilent 6890 N gas chromatograph-5973N mass detector system. The chromatograph was equipped with a HP-20M fused silica capillary column (50 m × 0.32 mm i.d., 0.3 µm film thickness). The splitless injector was maintained at 250°C. Helium was used as carrier gas (inlet pressure 10.5 psi) and the interface temperature was 230°C. The oven was held isothermally at 30°C for 1 min and then to 60°C at 30°C/min. The oven temperature was programmed from 60°C to 120°C at 3°C/min and then to 220°C at 2°C/min. 1-Heptanol (1 mg/L) was used as internal standard.

Statistical analyses

The volatile compounds were subjected to one factor analysis of variance (ANOVA, Statbox software). The results were considered significant if the associated *P* values were below 0.05.

Chapter V. Results

Identification and physiological characterization of strains

The 218 non-*Saccharomyces* yeasts stored in our collection were investigated. A total of 31 isolates were presumptively identified as *Hanseniaspora* and used to carry out this work (Table 2).

The ribosomal D1/D2 regions of these isolates were sequenced and, after screening for DNA homology using the BLAST program, yeast species identities were verified by comparing sequences of their ribosomal internal transcribed spacer region (Cadez *et al.,* 2002). We have found isolates belonging to four oenological species of the genus, *H. uvarum* (18), *H. vineae* (4), *H. osmophila* (1) and *H. guillermondii* (8) (Kurtzman & Robnett, 1998).

All studied strains were able to assimilate glucose for their growth (Table 3). Briefly, the API 20C AUX strips assay revealed that genus *Hanseniaspora* was not able to assimilate L-arabinose, inositol, lactose nor maltose. *H. osmophila* strain was not able to assimilate any of the other substrates included in the strips, under the assayed conditions while *H. vineae* strains were only able to assimilate cellobiose. All *H. guilliermondii* strains were able to assimilate this saccharyde and also 2-keto-glutarate and sucrose. Upon this global profile, there are strain specific differences for the other substrates, attributed to individual strain characteristics. According to RapID Yeast Plus strips (Data not shown), none of *Hanseniaspora* species used in this study had lipase, α-galactosidase, *β*-galactosidase, phosphatase, phosphatidylcoliesterase nor urease. Moreover, N-acetyl-glucosamine was not assimilated. As has been reported above some other minor differences were observed, probably due to strain rather than intraspecific differences. The use of these yeast identification tests, although in some cases is species-specific, has been used for a preliminary characterization of assimilation of carbon compounds. More accurated qualitative assays were further carried on to achieve a better characterization of biochemical activities.

Table 3. Results of the RapID Yeast Plus assays. (-) No assimilation, (+) Assimilation

Collection number	Species	Glucose	Maltose	Sucrose	Trehalose	Raffinose	Fatty acid ester	N-Acetyl-glucosamine	α-Glucoside	β-Glucoside	β-Galactoside	α-Galactoside	β-Fucoside	Phosphate	Phosphatidylcholine	Urea	Proline-β-naphthylamide	Histidine β-naphthylamide	Leucyl-glycine β-naphthylamide
A27	*H. guilliermondii*	+	+	+	+	+	-	-	+	+	-	-	+	-	-	-	-	+	+
A29	*H. guilliermondii*	+	-	-	-	-	-	-	+	+	-	-	-	-	-	-	+	+	+
B02	*H. guilliermondii*	+	+	+	+	+	-	-	+	+	-	-	+	-	-	-	-	+	+
B04	*H. guilliermondii*	+	-	-	-	-	-	-	+	+	-	-	+	-	-	-	-	+	+
C06	*H. guilliermondii*	+	+	+	-	+	-	-	+	+	-	-	+	-	-	-	-	+	+
D55	*H. guilliermondii*	+	+	+	+	+	-	-	+	+	-	-	+	-	-	-	-	+	+
D56	*H. guilliermondii*	+	-	-	-	-	-	-	+	+	-	-	+	-	-	-	-	+	+
D57	*H. guilliermondii*	+	-	-	-	-	-	-	+	+	-	-	-	-	-	-	+	+	+
C07	*H. osmophila*	+	+	+	+	-	-	-	+	+	-	-	-	-	-	-	-	+	+
E18	*H. uvarum*	+	+	-	-	-	-	-	-	-	-	-	-	-	-	-	+	+	-
E20	*H. uvarum*	+	+	-	-	-	-	-	-	-	-	-	-	-	-	-	+	+	-
E44	*H. uvarum*	+	-	+	-	+	-	-	+	-	-	-	-	-	-	-	+	+	+
F42	*H. uvarum*	+	+	+	+	+	-	-	-	+	-	-	-	-	-	-	-	+	-
G89	*H. uvarum*	+	-	-	-	-	-	-	+	+	-	-	-	-	-	-	-	+	-
G09	*H. uvarum*	+	-	+	-	+	-	-	-	-	-	-	-	-	-	-	-	+	-
H03	*H. uvarum*	+	+	-	-	-	-	-	-	-	-	-	-	-	-	-	+	+	+
H105	*H. uvarum*	+	-	-	-	+	-	-	-	+	-	-	-	-	-	-	-	-	-
H106	*H. uvarum*	+	-	-	-	+	-	-	-	-	-	-	-	-	-	-	-	-	-
H107	*H. uvarum*	+	-	-	-	-	-	-	-	+	-	-	-	-	-	-	-	+	-
H156	*H. uvarum*	+	-	-	-	-	-	-	+	+	-	-	-	-	-	-	-	+	-
H172	*H. uvarum*	+	-	+	-	+	-	-	+	-	-	-	-	-	-	-	+	+	+
B21	*H. uvarum*	+	+	+	+	-	-	-	-	+	-	-	-	-	-	-	-	+	-
B95	*H. uvarum*	+	-	-	-	+	-	-	-	+	-	-	-	-	-	-	-	-	-
M45	*H. uvarum*	+	-	+	-	+	-	-	+	-	-	-	-	-	-	-	+	+	+
M46	*H. uvarum*	+	+	-	-	+	-	-	-	-	-	-	-	-	-	-	+	+	+
M47	*H. uvarum*	+	-	-	-	+	-	-	-	+	-	-	-	-	-	-	-	-	-
M48	*H. uvarum*	+	-	-	-	+	-	-	-	-	-	-	-	-	-	-	-	-	-
E71	*H. vineae*	+	+	+	-	+	-	-	+	+	-	-	−	-	-	-	+	+	+
G26	*H. vineae*	+	-	-	-	-	-	-	+	+	-	-	+	-	-	-	-	+	+
P30	*H. vineae*	+	+	+	-	+	-	-	+	+	-	-	+	-	-	-	-	+	+
P38	*H. vineae*	+	+	+	-	+	-	-	+	+	-	-	+	-	-	-	+	+	+

Qualitative assays of enzymatic activities

The final results for the 31 *Hanseniaspora* strains characterized in this work are shown in table 4.

Table 4. Enzymatic activity of the *Hanseniaspora* strains

Yeast (N° isolates)	Pectinase			Esterase			Lipase			β-glucosidase			Polygalacturonase			Xylanase		
	−	+	++	−	+	++	−	+	++	−	+	++	−	+	++	−	+	++
H. uvarum (18)	1	6		6	6		8	10			6	12	14	4			1	3
	2															5		
H. vineae (4)	3	1		1	1		1	3				4	1	3		2	2	
H. guilliermondii (8)	6	2		3	3		8			3	1	3	5	3		6	2	
H. osmophila (1)	1			1	1		1				1	1	1					
Total (31)	2	9		1	20		1	13		3	8	20	21	10		2	7	
	2			1			8									4		

[a] Activity: (-), no activity; (+), moderate activity; (++), strong activity. Data obtained from triplicate assays

Similar patterns of overall enzyme activity were observed. Almost all strains displayed moderated or no lipase, esterase, polygalacturonase or pectinase activity. Moreover, a total of 28 strains showed β-glucosidase activity.

β-glucosidase and β-xylosidase activities

We have examined our isolates to determine β-glucosidase and β-xylosidase activities, the two more relevant glycosidic activities for non-*Saccharomyces* yeasts in wine (Mateo & Di Stefano, 1997; Fernández *et al.*, 2000; Strauss *et al.*, 2001; Mateo *et al.*, 2011). *H. uvarum* H107 and *H. vineae* (G26 and P38) exhibited strong reactions on both β-glucosidase and β-xylosidase quantitative assays (> 30 nkat) (Table 5).

These isolates were selected to quantify the level of production of these activities under enological stress conditions i.e. high sugar or ethanol concentrations.

Table 5. Glycolytic activities (nmol pNP h^{-1} 10^6 yeasts^{-1}=nkat) of the *Hanseniaspora* strains

Isolate	Species	β-Xylosidase (nkat)[a]	β-Glucosidase (nkat)[a]
A27	*H. guilliermondii*	2.65±0.08	1.70±0.17
A29	*H. guilliermondii*	2.18±0.34	8.09±0.22
B02	*H. guilliermondii*	2.38±0.07	21.07±0.11
B04	*H. guilliermondii*	2.24±0.84	4.83±0.23
C06	*H. guilliermondii*	4.33±0.03	5.27±0.21
D55	*H. guilliermondii*	3.60±0.64	36.10±0.13
D56	*H. guilliermondii*	2.99±0.14	20.80±2.23
D57	*H. guilliermondii*	2.72±0.45	6.66±0.26
C07	*H. osmophila*	2.92±0.23	5.91±0.89
E18	*H. uvarum*	2.31±0.57	3.06±0.84
E20	*H. uvarum*	2.11±0.17	42.49±0.43
E44	*H. uvarum*	2.99±0.62	2.99±0.11
F42	*H. uvarum*	2.86±0.11	3.26±0.56
G89	*H. uvarum*	1.90±0.26	2.92±0.19
G09	*H. uvarum*	9.45±0.23	40.38±0.64
H03	*H. uvarum*	1.90±0.07	2.58±0.05
H105	*H. uvarum*	3.60±0.71	5.64±0.34
H106	*H. uvarum*	15.83±3.83	16.36±3.16
H107	*H. uvarum*	38.34±0.14	48.74±2.83
H156	*H. uvarum*	2.58±0.23	5.30±0.23
H172	*H. uvarum*	2.24±0.78	21.89±2.98
B21	*H. uvarum*	12.11±0.13	8.06±0.32
B95	*H. uvarum*	2.99±0.64	20.08±0.11
M45	*H. uvarum*	22.86±0.12	4.63±0.24
M46	*H. uvarum*	1.90±0.29	3.26±0.82
M47	*H. uvarum*	18.45±0.21	2.91±0.39
M48	*H. uvarum*	18.17±0.14	5.24±0.24
E71	*H. vineae*	2.86±0.19	26.85±0.71
G26	*H. vineae*	39.22±0.84	48.81±0.78
P30	*H. vineae*	2.72±0.51	13.53±0.16
P38	*H. vineae*	31.00±0.71	51.26±3.11

[a] Average of duplicates

Effect of sugars or ethanol on β-glucosidase and β-xylosidase activities

The effects of sugar concentration (glucose, fructose and sucrose) on β-glucosidase and β-xylosidase were assayed. The β-glucosidase and β-xylosidase activities of the three strains were not affected by fructose (0-500 mM) and sucrose (0-50 mM) (Data not shown). The β-glucosidase activity from the three *Hanseniaspora* strains was slightly inhibited by low glucose concentrations (Fig. 1A).

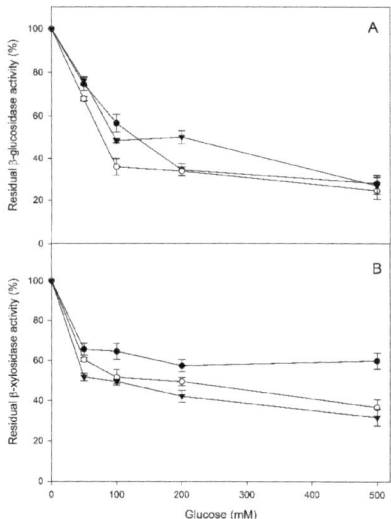

Figure 1. Effect of glucose concentration on glycolytic activities of different *Hanseniaspora* strains: (●) *H. uvarum* H107, (○) *H. vineae* G26, (▼) *H. vineae* P38. Enzyme activities were measured in the presence of various amounts of glucose under the standard assay conditions for A) β-glucosidase and B) β-xylosidase activities. Data represent means ± standard deviations from triplicate assays.

The enzyme remained 70-80% and 40-60% of relative activity in the presence of 50 and 100 mM glucose, respectively. Moreover, 30% remaining activity was detected in the presence of 500 mM glucose. Similar results have been observed in the three assayed strains. The β-xylosidase activity results were similar (Fig. 1B). We have also detected a moderated inhibition at 50 mM glucose, but 30-50 % activity was retained at higher glucose concentration (100-500 mM). The effect of ethanol concentration in range 0-20% (v/v) on both glycolytic activities was also determined. The β-glucosidase activity from the three *Hanseniaspora* strains was slightly inhibited by low and medium ethanol concentrations (0-10 %, v/v) (Fig. 2A).

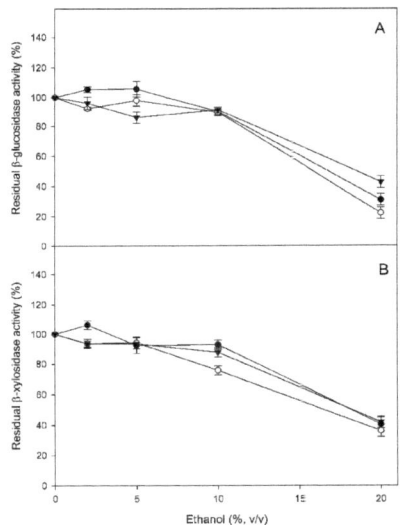

Figure 2. Effect of ethanol concentration on glycolytic activities of different *Hanseniaspora* strains. Legends are as shown in Figure 1.

In the presence of 20% (v/v) ethanol 20-50% activity remained. Similar results were expected of β-xylosidases (Fig. 2B). The remaining activity of the three strains at 20% (v/v) was similar (around 40%).

Effect of temperature and pH on β-glucosidase and β-xylosidase activities

The β-glucosidase activity did not vary with temperature ranging 4°C-60°C (Fig. 3A), with optimum of 30°C -40°C. However, β-xylosidase activity was highly influenced by temperature (Fig. 3B). Optimum temperature for the three strains was detected at 30°C, showing percentages of activity lower than 40% at extreme values (4°C-60°C).

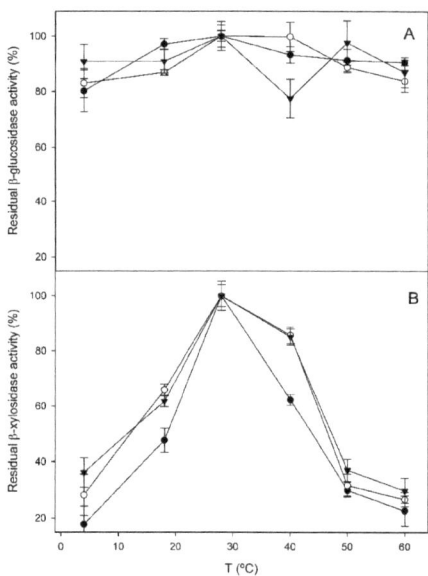

Figure 3. Effect of temperature on glycolytic activities of different *Hanseniaspora* strains. Legends are as shown in Figure 1.

The pH optimum was also evaluated in phosphate buffer ranging from pH 3.0 to 8.0. The β-glucosidase activity from the three *Hanseniaspora* strains was slightly influenced inside the interval assay, with residual activities between 80% - 100% (Fig. 4A). β-Xylosidase exhibited high pH stability between pH 5.0 to 8.0, retaining more than 80% residual activity *in vitro*, with an optimum value around pH 6.0-7.0 (Fig. 4B).

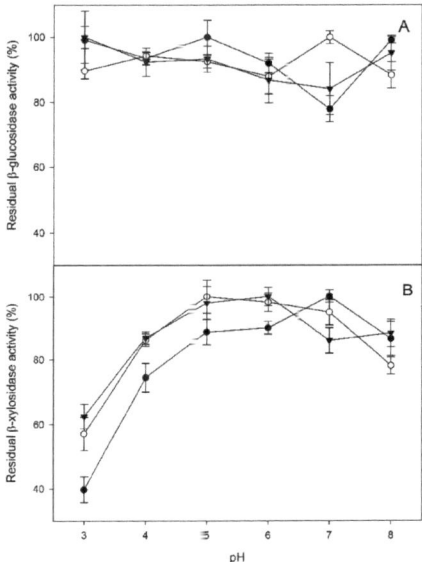

Figure 4. Effect of pH on glycolytic activities of different *Hanseniaspora* strains. Legends are as shown in Figure 1.

Screening and quantitation of protease activity

Once all isolates were identified, they were inoculated on YPD casein agar plates to detect the production of exocellular protease; only six *Hanseniaspora* isolates showed this enzymatic activity and were selected to continue the assay At a first step, quantitative procedure for protease activity was reviewed and the use of a 0. M Tris-HCl pH 7.5 buffer or 0.1 M citrate phosphate pH 6.0 buffer was assayed. As can be seen, the most promising data were obtained with citrate-phosphate buffer which was used routinely in further assays (Table 6). In the same way, some differences were detected depending on the substrate used to quantify protease activity (casein or BSA) (Table 6). As can be seen, the most promising results were obtained when substrate was BSA. This result could be explained because of the low solubility of casein at pH values used in the quantitative procedure used in this work. Assays with 0.1 M citrate phosphate buffer pH 6.0 buffer and using BSA as a substrate were routinely used in this work.

Table 6. Effect of buffer and substrate on protease activity.

Hanseniaspora isolate	0.1 M Tris-HCl pH 7.5		0.1 M citrate phosphate pH 6.0	
	Casein	BSA	Casein	BSA
B04	0.008 ± 0.002	0.030 ± 0.005	0.006 ± 0.002	0.407 ± 0.022
D56	0.025 ± 0.004	0.058 ± 0.008	0.069 ± 0.008	0.279 ± 0.014
D57	0.021 ± 0.003	0.047 ± 0.006	0.017 ± 0.003	0.405 ± 0.018
E71	0.012 ± 0.002	0.006 ± 0.002	0.005 ± 0.002	0.158 ± 0.011
A27	0.009 ± 0.002	0.054 ± 0.004	0.005 ± 0.002	0.317 ± 0.021
C07	0.110 ± 0.002	0.045 ± 0.005	0.009 ± 0.003	0.237 ± 0.015

Data represent means ± standard deviations from triplicate assays

Effect of sugars and ethanol on protease activity

The effects of sugar concentration (glucose or fructose) on protease activity were also assayed (Figure 5). Low sugar concentrations slightly increase protease activity so that, except for B04 isolate, maximum values were obtained with 100 mM sugar in the medium; this increase in the activity is higher with the presence of fructose. Enzymatic activity decreased with higher sugar concentrations but 40-60% remaining activity was detected in the presence of 500 mM glucose; B57 and A27 isolates maintain 80 % of activity in the presence of 500 mM of fructose.

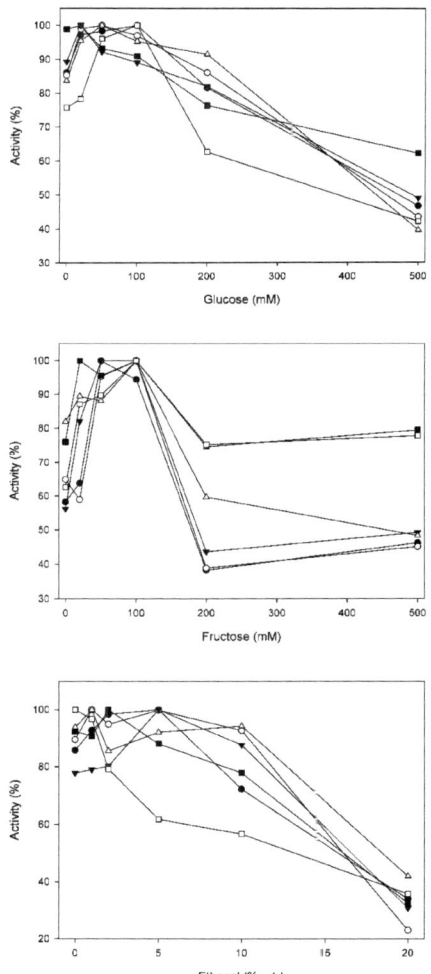

Figure 5. Effect of glucose, fructose and ethanol concentration on proteolytic activity of different *Hanseniaspora* strains: ■ B04, Δ B56, ▼ A27, ● C07, □ B57, ○E71.

Sugar content affects the water activity of the medium and also, the water availability in the active site of the enzyme. As proteases are hydrolase enzymes, a lower concentration of water can affect the enzymatic reaction, diminishing the synthesis of products.

The effect of ethanol concentration on protease was also assayed (Figure 5). As can be observed, B57 isolate showed a different evolution, so activity decreased almost 40% with 5% ethanol. Other isolates showed higher activities with this ethanol content but only enzyme from C07 and B56 isolates maintain about 100% activity. For all isolates, protease activity decreased more than 50% with 20% ethanol content in the medium. The change in polarity of the medium induced by ethanol could alter enzyme conformation and, consequently, its active site, thus reducing its activity. Nevertheless, the enzymatic activities were unaffected by such phenomenon; this may due to both the intrinsic structural characteristics of the enzyme and/or active site and a probable protective mechanism that reduces the unfolding rate due to immobilisation of the enzyme on the yeast cell membranes (Barbagallo *et al.*, 2003). Alternatively, higher ethanol concentrations may have altered membrane permeability thereby allowing easier access between the intracellular enzyme and substrate (Pemberton *et al.*, 1980).

Effect of temperature and pH on protease activity

The optimum pH was evaluated in phosphate buffer ranging from pH 3.0 to 8.0. The protease activity from the *Hanseniaspora* strains was influenced inside the interval assay, with an optimum value around pH 6 (Figure 6). As can be seen, only a residual activity can be detected at acidic pH values. Protease activity was strongly influenced by temperature (Figure 6). Optimum temperature for all isolates was detected at 50 °C, showing percentages of activity lower than 40% at extreme values (4°C-60°C).

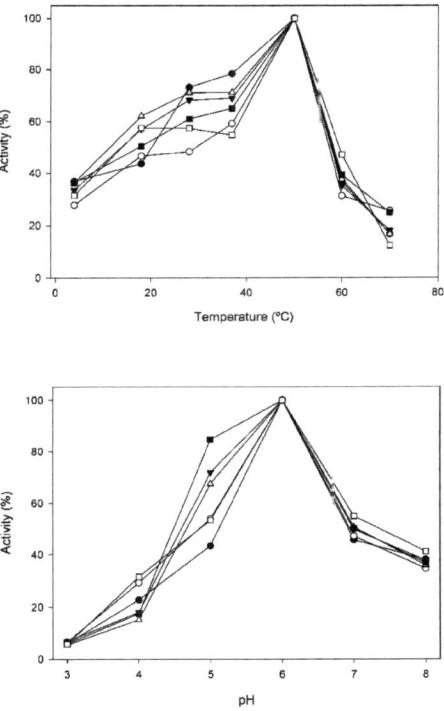

Figure 6. Effect of temperature and pH on proteolytic activity of different *Hanseniaspora* strains. Legends are as shown in Figure 5.

Determination of volatile compounds liberated from wine treated with different *Hanseniaspora* yeasts

Muscat juice was used for vinification with a commercial *S. cerevisiae* strain. Afterwards, *H. uvarum* H107 and *H. vineae* G26 and P38 were inoculated (in triplicate assays) and terpene and other volatile compounds were determined (Table 7).

Table 7. Terpene and other volatile compounds in Muscat wine. Concentration expressed as $\mu g/L^a$.

	Control[b]	Hanseniaspora inoculated		
		H. uvarum H107	H. vineae G26	H. vineae P38
Oxide A[c]	29.7 (1.2)	30.4 (2.1)	33.7 (3.2)	26.9 (3.4)
Oxide B[d]	nd	nd	nd	nd
Linalool	20.0 (0.9)	40.4*(3.9)	47.4*(3.4)	38.2*(5.3)
Ho-trienol	24.0 (3.2)	51.3*(5.3)	35.1*(4.2)	24.9*(0.6)
2-Phenylethanol	1890.2 (43.4)	3057.5*(39.8)	2747.8*(26.8)	2568.5*(45.6)
Oxide C[e]	nd	nd	nd	nd
Oxide D[f]	nd	nd	nd	nd
Terpineol	53.3 (3.4)	67.2*(4.7)	65.1*(1.2)	54.5 (3.9)
Nerol	24.6 (2.8)	25.8 (1.1)	23.4 (3.1)	26.3 (1.2)
Geraniol	59.8 (5.0)	61.3 (3.7)	56.9 (1.7)	62.8 (1.7)
Diol 1[g]	43.2 (4.7)	87.9*(2.1)	80.2*(2.1)	81.2*(3.2)
4-Vinylphenol	63.2 (1.2)	89.7*(2.4)	75.7*(5.8)	62.1 (0.9)
Endiol[h]	nd	58.8*(2.1)	52.0*(3.4)	34.1*(4.2)
Diol 2[i]	12.0 (0.6)	13.4 (0.9)	7.8 (2.6)	10.1 (0.9)
2-Phenylethyl acetate	28.0 (4.1)	56.2*(7.2)	23.3 (1.2)	25.8 (4.7)
2-Methoxy-4-vinylphenol	89.0 (6.1)	103.0*(5.3)	105.4*(6.5)	94.1 (2.9)

[a] Values in brackets represent standard deviation (n=3). ANOVA one factor, significant difference is indicated as * (p<0.05).

[b] Wine produced only with *Saccharomyces cerevisiae*.

[c] *cis*-5-vinyltetrahydro-1,1,5-trimethyl-2-furanmethanol.

[d] *trans*-5-vinyltetrahydro-1,1,5-trimethyl-2-furanmethanol.

[e] *cis*-6-vinyltetrahydro-2,2,6-trimethyl-2H-pyran-3-ol.

[f] *trans*-6-vinyltetrahydro-2,2,6-trimethyl-2H-pyran-3-ol.

[g] 2,6-Dimethyl-3,7-octadien-2,6-diol.

[h] 2,6-Dimethyl-7-octene-2,6-diol.

[i] 2,6-Dimethyl-2,7-octadien-1,6-diol.

nd: not detected

We were not able to detect significant increase in the level of nerol and geraniol after the addition of *Hanseniaspora* isolates. This can be attributed to the fact that the production of these compounds could be previously carried out by *Saccharomyces* strains during the alcoholic fermentation of Muscat wines (Mateo & Di Stefano, 1997; Restuccia *et al.*, 2002). The concentration of *cis*-5-vinyltetrahydro-1,1,5-trimethyl-2-furanmethanol was not increased despite of the addition of *Hanseniaspora*, while as the other oxides (*trans*-5-vinyltetrahydro-1,1,5-trimethyl-2-furanmethanol, *cis*-6-vinyltetrahydro-2,2,6-trimethyl-2H-pyran-3-ol and *trans*-6-vinyltetrahydro-2,2,6-trimethyl-2H-pyran-3-ol) were not detected. Morever, 2,6-dimethyl-2,7-octadien-1,6-diol levels were not affected.

Linalool and derivative compounds and aromatic alcohols were detected at the same level of control wine, after the addition of *Hanseniaspora* isolates. Terpineol increased only after the addition of *H. uvarum* H107 or *H. vineae* G26, but not when *H. vineae* P38 was added. The same results were recorded for 4-vinyl-phenol and 2-methoxy-4-vinylphenol. Linalool, 2-phenyl ethanol and 4-vinyl-phenol compounds are associated with fruity characteristics (Maicas & Mateo, 2005; López de Lerma & Peinado, 2011).

The analysis of the other compounds revealed an increase in concentration when yeasts were added, therefore assessing the effect of glycosidases. Wines treated with *Hanseniaspora* showed higher concentrations of 2-phenyl ethanol (Gunata *et al.*, 1988; Dubordieu *et al.*, 1989). *H. uvarum* H107 provides improvements in 2-phenylethyl acetate production as previously reported for other *Hanseniaspora* strains (Rojas *et al.*, 2001; Moreira *et al.*, 2005; Viana *et al.*, 2009).

Chapter VI. Discussion

The criteria for discrimination of yeast species proposed by Kurtzman and Robnett (1998) using D1/D2 rDNA sequencing was used in this work: a match greater than 99% was required to discriminate among yeast species. The analysis of the miniaturized biochemical assay strips basically confirmed these results, although some other minor differences were observed probably due to strain rather than intraspecific differences. The use of these yeast identification tests, although in some cases is species-specific, can be used for a preliminary characterization of assimilation of carbon compounds (strain typing), rather than for identification purposes.

The role of microbial glycosidases on the hydrolysis of glycosides during the winemaking process contribution was detected. From a technological point of view, this activity justifies the utilisation of *Hanseniaspora* in industrial production of aromatic wines (Fernández *et al.*, 2003; Mateo *et al.*, 2011; Swangkeaw *et al.*, 2011). However, yeasts belonging to the *Hanseniaspora* genus have been considered as spoilage yeasts, particularly during the first stage of fermentation, due to the undesirable production of componds such as acetic acetic acid or ethyl acetate (Ciani & Comitini, 2011). The significance of glycosidases for the wine industry lies in their potential for releasing flavor compounds from glycosidically bound nonvolatile precursors in wine (Gueguen *et al.*, 1997a; Ubeda & Briones, 2000; Ferreira *et al.*, 2001; Strauss *et al.*, 2001; Maicas & Mateo, 2005; Arévalo *et al.*, 2007b). Some aromatic precursors (terpenes) found in wine are not volatile and therefore they cannot contribute to wine aroma. *Hanseniaspora* strains included in this work showed β-glucosidase and β-xylosidase activities (remarkable *H. uvarum* H107 and *H. vineae* G26 and P38). To elucidate if wine conditions could affect their activity, as suggested by Grimaldi *et al.* (2005), we have also studied the influence of sugar and ethanol addition to simulate native conditions in wine. Our results were in agreement with Strauss *et al.* (2001), detecting a significant negative influence at high glucose concentration on the activities of β-glucosidase and β-xylosidase. This negative influence was greater than that exerted by the ethanol on the same enzymes, suggesting that they may be less effective in wine. These results also confirm data previously reported by our group in other non-*Saccharomyces* strains (Mateo *et al.*, 2011; Romero *et al.*, 2012). Our report is surprising as *Hanseniaspora* is a genus described as ethanol inhibited *in vivo* (Branco *et al.*, 2012). It has been previously reported that low quantities of alcohol stimulate β-glucosidase activity (Swangkeav *et al.*, 2011). These results are similar to

46

our findings for β-glucosidase and mainly β-xylosidase. These authors argued that the glycosyltransferase activity of the enzyme could be conditioning this fact.

The β-glucosidase activity was not affected by temperature while β-xylosidase activity was highly influenced. The first result is very important because winemaking is nowadays conducted at low temperatures, so that it is important to obtain enzymes with high activity under these conditions. The high activity of β-glucosidase activity at different values of pH also permits a broad interval of use. However the β-xylosidase optimal range of use was at pH 5.0 to 8.0. The residual activity, even at low pH (3.0 to 5.0), was good enough to permit their use in wine conditions.

Yeast protease may liberate amino acids and peptides from grape protein during fermentation which can benefit growth of microorganisms during or after alcoholic fermentation. Another aspect is that yeast cells may release nitrogen containing metabolites to the media. The composition of amino acids peptides and proteins in wine is based on grape related compounds transferred and transformed during the winemaking process and breakdown products through the protease activity from yeasts and compounds released by yeasts (Alexandre *et al.*, 2001). Results obtained in our laboratory in previous work allow to conclude that protease activity in *Pichia* and *Wickerhamomyces* isolates was very low (Madrigal *et al.*, 2013), according with results obtained by other authors (Charoenchai *et al.*, 1997; Strauss *et al.*, 2001). These authors suggested that *Hanseniaspora* isolates could be interesting to obtain this enzymatic activity, but some contradictory data have been obtained. Many of these studies have been conducted with *H. uvarum* (*K. apiculata*) isolates and, on the basis of the results obtained in our work, exocellular protease this specie has a very low activity. On the other hand, assays made by these authors have used acidic pH buffers and we have shown that protease from *Hanseniaspora* yeasts is pH dependent, showing maximum values at pH 6.0.

The aspartic proteases secreted by non-*Saccharomyces* yeasts have a tertiary structure consisting of two approximately symmetric lobes with each lobe carrying an aspartic acid residue to form the catalytic site. Unlike the other types of proteases, the activity of the aspartic proteases is dependent on pH conditions (Borrelli *et al.*, 2008; Cascella *et al.*, 2005). Aspartic endopeptidases (E3.4.23.x) are widely distributed in living organisms from vertebrates to fungi, plants and retroviruses. Most of these enzymes are composed of approximately 323 to 340 amino acid residues, with molecular weights ranging between 35 000 to 50 000 Daltons (Da) and isoelectric points (pI) ranging between 3 and 4.5 because of the high percentage of acidic amino acid residues (about 13%) in the proteins. They have optimum function at pH 3 to 4. They show substrate specificity towards extended peptide substrates and residues with large hydrophobic side chains on either side of the scissile bond (Rawlings *et al.*, 2009).

Nevertheless, according to the MEROPS and Protein Data Bank (PDB), there are eight sub-families within the aspartic proteases. These subfamilies differ according to the specific residues in the active site, the position of the catalytic aspartic acid residues in the peptide chains, substrate specificity, the number of disulfide bridges in their structure and the optimal pH at which the enzymes function, varying from acidic to neutral (Rawlings & Bateman, 2009; Rawlings *et al.*, 2011).

Certain proteases have been used in food processing for centuries. Rennet (mainly chymosin), obtained from the fourth stomach (abomasum) of unweaned calves has been used traditionally in the production of cheese (Ozcan & Kurdal, 2012). Similarly, papain from the leaves and unripe fruit of the pawpaw (*Carica papaya*) has been used to tenderise meats. Proteases may be used at various pH values, and they may be highly specific in their choice of cleavable peptide links or quite non-specific. Proteolysis generally increases the solubility of proteins at their isoelectric points (Ozcan & Kurdal, 2012).

The action of rennet in cheese making is an example of the hydrolysis of a specific peptide linkage, between phenylalanine and methionine residues ($-Phe_{105}-Met_{106}-$) in the K-casein protein present in milk. The major problem that had to be overcome in the development of the microbial rennets was temperature lability. Chymosin is a relatively unstable enzyme and once it has done its major job, little activity remains. However, the enzyme from *Mucor miehei* retains activity during the maturation stages of cheese-making and produces bitter off-flavours. The development of unwanted bitterness in ripening cheese is an example of the role of proteases in flavour production in foodstuffs. The presence of proteases during the ripening of cheeses is not totally undesirable and a protease from *Bacillus amyloliquefaciens* may be used to promote flavour production in cheddar cheese. Lipases from *M. miehei* or *Aspergillus niger* are sometimes used to give stronger flavours in Italian cheeses by a modest lipolysis, increasing the amount of free butyric acid. They are added to the milk (30 U/L[1]) before the addition of the rennet (Ozcan & Kurdal, 2012).

The action of endogenous proteases in meat after slaughter is complex but 'hanging' meat allows flavour to develop, in addition to tenderising it. It has been found that peptides with terminal acidic amino acid residues give meaty, appetising flavours akin to that of monosodium glutamate. Non-terminal hydrophobic amino acid residues in medium-sized oligopeptides give bitter flavours, the bitterness being less intense with smaller peptides and disappearing altogether with larger peptides. Application of this knowledge allows the tailoring of the flavour of protein hydrolysates (Ha *et al.*, 2013)

Correctly applied proteolysis of inexpensive materials such as soya protein can increase the range and value of their usage, as indeed occurs naturally in the production of soy sauce. Partial hydrolysis of soya protein greatly increases its 'whipping expansion', further hydrolysis improves its emulsifying capacity. If their flavours are correct, soya protein hydrolysates may be added to cured meats. Hydrolysed proteins may develop properties that contribute to the elusive, but valuable, phenomenon of 'mouth feel' in soft drinks (Hrčková *et al.*, 2002)

Proteases are also used in the baking industry. Where appropriate, dough may be prepared more quickly if its gluten is partially hydrolysed. A heat-labile fungal protease is used so that it is inactivated early in the subsequent baking. Weak-gluten flour is required for biscuits in order that the dough can be spread thinly and retain decorative impressions. In the past this has been obtained from European domestic wheat but this is being replaced by high-gluten varieties of wheat. The gluten in the flour derived from these must be extensively degraded if such flour is to be used efficiently for making biscuits or for preventing shrinkage of commercial pie pastry away from their aluminum dishes (Renzettia & Arendta, 2009).

Monoterpenes, norisoprenoids, benzene derivatives and aliphatic components are traditionally involved in Muscat grape juice and wine. These compounds have been detected in glycosidically-bound form: therefore, the liberation could enhance wine aroma. In order to confirm our previous laboratory results, assays were also carried out in Muscat wine, and volatile compounds were analysed by GC/MS. Muscat wine (13% v/v initial alcohol) showed only a moderated overall terpene increase (1.1-1.3 folds) when treated with these strains. These results are conditioned by the effect of ethanol on glycolytic enzymes. Going into detail, the use of these strains offered an increase of the levels of ho-trienol, 2-phenylethanol and 2,6-dimethyl-3,7-octadien-2,6-diol in wine. The sum of ho-trienol, linalool and terpineol seems to play an important role in the aromatic definition of the wines of Loureiro and Alvarinho varieties (Oliveira *et al.,* 2008). 2-Phenylethanol also participates to confer fruity and floral notes to these wines, and its presence is related to the metabolic activity of the non-*Saccharomyces* yeasts (Swiegers *et al.,* 2005). Our findings are similar to the observations of Fernandez *et al.* (2003) who shown the ability of several wine yeasts to hydrolyze terpenoids, norisoprenoids and benzenoids glycosides; among wine yeasts *H. uvarum* was able to hydrolyse both glycoconjugated forms of pyranic and furanic oxides of linalool. Our results open the possibility to the use of these strains to be used to improve the aromatic characteristics of the wines, in regard to liberation of terpenes. The production of wines with the addition of non-*Saccharomyces* strains has been traditionally related to high concentrations in vinyl-phenols (4-vinyl-phenol, 4-vinyl-guayacol) reaching concentrations up to 1 mg/L (Sefton & Williams, 1991; Gunata, 1993). The concentration of 4-vinyl-phenol in the tested wines was under 90 μg/L, which enables the use of our selected strains in winemaking.

Conclusions

In the present study we have characterized different *Hanseniaspora* strains, previously isolated and stored in our laboratory. The strains have been classified according to the molecular results and have also been assayed to determine some biochemical activities of oenological interest. The yeasts basically showed moderated or no lipase, esterase, polygalacturonase, pectinase or xylanase activity but interestingly glycolitic (β-xylosidase and β-glucosidase) activities were detected. These enzymatic activities can be used to enhance the quality of wine produced with *S. cerevisiae* strains.

This study contributes to the research on the role and potential exploitation of *Hanseniaspora* strains to release of volatile terpenes in wines. The potential enzymatic activities showed by these isolates could be profitable used in specific vinification processes by consuming the remaining precursors of aromatic compounds. The two wineries supporting this study have decided to evaluate the use of the selected strains in further campaigns.

Bibliography

Alexandre, H., Heintz, D., Chassagne, D., Guilloux–Benatier, M., Charpentier, C. & Feuillat, M., 2001. Protease A activity and nitrogen fractions released during alcoholic fermentation and autolysis in enological conditions. J. Ind. Microbiol. Biotechnol. 26, 235–240.

Andorrà, I., Landi, S., Mas, A., Guillamón, J.M. & Esteve–Zarzoso, B., 2008. Effect of oenological practices on microbial populations using culture–independent techniques. Food Microbiol. 25, 849–856.

Andorrà, I., Berradre, M., Rozès, N., Mas, A., Guillamon, J.M. & Esteve–Zarzoso, B., 2010. Effect of pure and mixed cultures of the main yeast species on grape must fermentation. European J. Food Res. Technol. 231, 215–224.

Anfang, N., Brajkovich, M. & Goddard, M.R., 2009. Co–fermentation with *Pichia kluyveri* increases varietal thiol concentrations in Sauvignon blanc. Aust. J. Grape Wine Res. 15, 1–8.

Arévalo, M., Úbeda, J.F. & Briones, A.I., 2007a. Glucosidase activity in wine yeasts: Application in enology. Enzyme Microb. Technol. 40, 420–425.

Arevalo–Villena, M., Ubeda Iranzo, J. & Briones Perez, A., 2007b. Enhancement of aroma in white wines using a beta–glucosidase preparation from *Debaryomyces pseudopolymorphus* (A–77). Food Biotechnol. 21, 181–194.

Arroyo–López, F.N., Durán–Quintana, M.C., Ruiz–Barba, J.L., Querol, A. & Garrido–Fernández, A., 2006. Use of molecular methods for the identification of yeast associated with table olives. Food Microbiol. 23, 791–796.

Arroyo–López, F.N., Querol, A., Bautista Gallego, J. & Garrido Fernández, A., 2008. Role of yeasts in table olive production. Int. J. Food Microbiol. 128, 189–196.

Barata, A., González, S., Malfeito–Ferreira, M., Querol, A. & Loureiro, V., 2008a. Sour rot damaged grapes are sources of wine spoilage yeasts. FEMS Yeast Res. 8, 1008–1017.

Barata, A., Seborro, F., Belloch, C., Malfeito–Ferreira, M. & Loureiro, V., 2008b. Ascomycetous yeast species recovered from grapes damaged by honeydew and sour rot. J. Appl. Microbiol. 104, 1182–1191.

Barata, A., Campo, E., Malfeito–Ferreira, M., Loureiro, V.l., Cacho, J. & Ferreira, V., 2011. Analytical and sensorial characterization of the aroma of wines produced with sour rotten grapes using GC–O and GC–MS: identification of key aroma compounds. J. Agric. Food Chem. 59, 2543–2553.

Barbagallo, R.N., Spagna, G., Palmeri, R., Restuccia, C. & Giudici, P., 2004. Selection, characterization and comparison of □-glucosidase from moulds and yeasts employable for enological applications. Enz. Microb. Technol. 35, 58-66.

Barrett, A.J., Rawlings, N.D. & Woessner, J.F., 2004 Handbook of Proteolytic Enzymes, second ed. Academic Press, London.

Belin, J.M., 1972. Recherches sur la répartition des levures à la surface de la grappe de raisin. Vitis 11, 135–145.

Borelli, C., Ruge, E., Lee, J.H., Schaller, M., Vogelsang, A., Monod, M., Korting, H.C., Huber, R. & Maskos, K., 2008. X–ray structures of Sap1 and Sap5: Structural comparison of the secreted aspartic proteinases from *Candida albicans*. Proteins: Structure, Functions and Bioinformatics 72, 1308–1319.

Boundy–Mills, K., 2006. Methods for investigating yeast biodiversity. In: Péter, G., Rosa, C. (Eds.), Biodiversity and Ecophysiology of Yeasts. Springer, Berlin Heidelberg, pp. 67–100.

Branco, P., Monteiro, M., Moura, P. & Albergaria, H., 2012. Survival rate of wine–related yeasts during alcoholic fermentation assessed by direct live/dead staining combined with fluorescence in situ hybridization. Int. J. Food Microbiol. 158, 49–57.

Brimer, L., Nout, M.J.R. & Tuncel, G., 1998. Glycosidase (amygdalase and linamarase) from *Endomyces fibuliger* (LU677): formation and crude enzyme properties Appl. Microbiol. Biotechnol. 49, 182–188.

Cadez N., Raspor, P., de Cock, A.W.A.M., Boekhout T. & Smith, M.T., 2002. Molecular identificacion and genetic diversity whitin species of genera *Hanseniaspora* and *Kloeckera*. FEMS Yeast Res. 1, 279–289.

Cadez, N., Poot, G.A., Raspor, P. & Smith, M.T., 2003. *Hanseniaspora meyeri* sp. nov., *Hanseniaspora clermontiae* sp. nov., *Hanseniaspora lachancei* sp. nov. and *Hanseniaspora opuntiae* sp. nov., novel apiculate yeast species. Int. J. Syst. Evol. Microbiol. 53, 1671–1680.

Cascella, M., Micheletti, C., Rothlisberger, U. & Carloni, P., 2005. Evolutionarity conserved functional mechanics across pepsin–like and retroviral aspartic proteases. J. Am. Chem. Soc. 127, 3734–3742.

Chairote, G., Rodriguez, F. & Crouzet, J., 1981. Characterization of additional volatile flavor components of apricot. J. Agric. Food Chem. 46, 1898–1901.

Charoenchai, C., Fleet, G.H., Henschke, P.A. & Todd, B.E.N., 1997. Screening of non–Saccharomyces wine yeasts for the presence of extracellular hydrolytic enzymes. Austr. J. Grape & Wine Res. 3, 2–8.

Ciani, M. & Maccarelli, F., 1998. Oenological properties of non–Saccharomyces yeasts associated with wine–making. World J. Microbiol. Biotechnol. 14, 199–203.

Ciani, M., Fatichenti, F. & Mannazzu, I., 2002. Yeasts in winemaking biotechnology. In: Ciani, M. (Ed.), Biodiversity and Biotechnology of Wine Yeasts. Research Signpost, Kerala, India, pp. 112–123.

Ciani, M., Comitini, F., Mannazzu, I. & Domizio, P., 2010. Controlled mixed culture fermentation: a new perspective on the use of non–Saccharomyces yeasts in winemaking. FEMS Yeast Res. 10, 123–333.

Ciani, M. & Comitini, F., 2011. Non–Saccharomyces wine yeasts have a promising role in biotechnological approaches to winemaking. Ann. Microbiol. 61, 25–32.

Clemente–Jimenez, J.F., Mingorance–Cazorla, L., Martinez–Rodriguez, S., Las Heras–Vazquez, F.J. & Rodriguez–Vico, F., 2004. Molecular characterization and oenological properties of wine yeasts isolated during spontaneous fermentation of six varieties of grape must. Food Microbiol. 21, 149–155.

Cocolin, L., Campolongo, S., Alessandria, V., Dolci, P. & Rantsiou, K., 2011. Culture independent analyses and wine fermentation: an overview of achievements 10 years after first application. Annals Microbiol. 61, 17–23.

Combina, M., Mercado, L., Borgo, P., Elia, A., Jofré, V., Ganga, A., Martinez, C. & Catania, C., 2005. Yeasts associated to Malbec grape berries from Mendoza, Argentina. J. Appl. Microbiol. 98, 1055–1061.

Comitini, F., Gobbi, M., Domizio, P., Romani, C., Lencioni, L., Mannazzu, I. & Ciani, M., 2011. Selected non–*Saccharomyces* wine yeasts in controlled multistarter fermentations with *Saccharomyces cerevisiae*. Food Microbiol. 28, 873–882.

Cordero Otero, R., Ubeda, J.F., Briones–Perez, A.I., Potgieter, N., Villena, M.A., Pretorius, I.S. & Van Rensburg, P., 2003. Characterization of the β–glucosidase activity produced by enological strains of non–*Saccharomyces* yeast. J. Food Sci. 68, 2564–2569.

Cordonnier, R. & Bayonove, C., 1974. Mise en evidence dans la baie de raisin, variété Muscat d'Alexandrie, de monotèrpenes liés, révélables par une ou plusieurs encimes du fruti. C. R. Acad. Sci. Paris 278, 3387– 3390.

Cramer, A.C., Vlassides, S. & Block, D.E., 2002. Kinetic model for nitrogen–limited wine fermentations. Biotechnol. Bioeng. 77, 49–60.

Cray, J.A., Bell, A.N.W., Bhaganna, P., Mswaka, A.Y., Timson, D.J. & Hallsworth, J.E., 2013. The biology of habitat dominance; can microbes behave as weeds? Microb. Biotechnol. 6, 453–492.

Darriet, P., Boidron, J.N. & Dubourdieu, D., 1988. L'hydrolyse des hétérosides terpéniques du Muscat a Petit Grains par les enzymes périplasmiques de *Saccharomyces cerevisiae*. Conn. Vigne Vin 22, 189–195.

de Llanos, R., Fernández–Espinar, M.T. & Querol, A., 2004. Identification of species of the genus *Candida* by analysis of the 5.8S rRNA gene and the two ribosomal internal transcribed spacers. Antonie van Leeuwenhoek 85, 175–185.

Deák, T. & Beuchat, L.R., 1996. Handbook of Food Spoilage Yeasts CRC Press. Boca Raton, Florida.

Delcroix, A., Gunata, Z., Sapis, J.C., Salmon, J.M. & Bayonove, C., 1994. Glycosidase activities of three enological yeast strains during wine making. Effect on the terpenol content of Muscat wine. Am. J. Enol. Vitic. 45, 291–296.

Di Maio, S., Genna, G., Gandolfo, V., Amore, G., Ciaccio, M. & Oliva, D., 2012. Presence of *Candida zemplinina* in Sicilian musts and selection of a strain for wine mixed fermentations. S. Afr. J. Enol. Vitic. 33, 80–87.

Di Maro, E., Ercolini, D. & Coppola, S., 2007. Yeast dynamics during spontaneous wine fermentation of the Catalanesca grape. Int. J. Food Microbiol. 117, 201–210.

Di Stefano, R., Magiorotto, G. & Gianotti, S., 1992. Transformazioni di nerolo e geraniolo indotte dai lieviti. Riv. Vitic. Enol. 42, 43–49.

Dizy, M. & Bisson, L.F., 2000. Proteolytic activity of yeast stains during grape juice fermentation. Am. J. Enol. Vitic. 51, 155–167.

Domizio, P., Romani, C., Lencioni, L., Comitini, F., Gobbi, M., Mannazzu, I. & Ciani, M., 2011. Outlining a future for non–*Saccharomyces* yeasts: selection of putative spoilage wine strains to be used in association with *Saccharomyces cerevisiae* for grape juice fermentation. Int. J. Food Microbiol. 147, 170–180.

Dubordieu, D.P., Darriet, P., Chatonnet, F. & Boidron, J.M., 1989. Proceedings of the Fourth International Enology Symposium, 151–159.

Dupin, I., Gunata, Z., Sapis, J.C., Bayonove, C., M'Bairaroua, O. & Tapiero, C., 1992. Production of a □–apiosidase by *Aspergillus niger*. Partial purification, properties and effect on terpenyl apiosylglucosides from grape. J. Agric. Food Chem. 40, 1886–1891.

Engel, K.H., Flath, R.A., Butter, R.G., Mon, T.R., Ramming, D.W. & Teranishi, R., 1988. Investigation of volatile constituents in nectarines. 1. Analytical and sensory characterization of aroma components in some nectarine cultivars. J. Agric. Food Chem. 36, 549–553.

Erten, H. & Campbell, I. (2001) The production of low–alcohol wines by aerobic yeasts. J Inst Brew 107: 207–215.

Esteve–Zarzoso, B., Belloch, C., Uruburu, F. & Querol, A., 1999. Identification of yeasts by RFLP analysis of the 5.85 rRNA gene and the two ribosomal internal transcribed spacers. Int. J. Syst. Bacteriol. 49, 329–337.

Fernández, M., Úbeda, J.F. & Briones, A.I., 2000. Typing of non–*Saccharomyces* yeasts with enzymatic activities of interest in wine–making. Int. J. Food Microbiol. 59, 29–36.

Fernandez, M., Di Stefano, R. & Briones, A., 2003. Hydrolysis and transformation of terpene glycosides from Muscat must by different yeast species. Food Microbiol. 20, 35–41.

Fernández–Espinar, M., Esteve–Zarzoso, B., Querol, A. & Barrio, E., 2000. RFLP analysis of the ribosomal internal transcribed spacers and the 5.8S rRNA gene region of the genus *Saccharomyces*: a fast method for species identification and the differentiation of flour yeasts. Antonie van Leeuwenhoek 78, 87–97.

Fernández–Espinar, M.T., Llopis, S., Querol, A. & Barrio, E., 2011. Molecular identification and characterization of wine yeasts, In: Carrascosa, A.V., Muñoz, R., González, R. (Eds.), Molecular Wine Microbiology, 1st ed. Elsevier, Academic Press, London, UK, pp. 111–141.

Ferreira, A.M., Clímaco, M.C. & Faría, A.M., 2001. The role of non–*Saccharomyces* species in releasing glycosidic bound fraction of grape aroma components – A preliminary study. J. Appl. Microbiol. 91, 67–71.

Fleet, G.H., Lafon–Lafourcade, S. & Ribéreau–Gayon, P., 1984. Evolution of yeasts and lactic acid bacteria during fermentation and storage of Bordeaux Wines. Appl. Environ. Microbiol. 48, 1034–1038.

Fleet, G., Prakitchaiwattana, C., Beh, A. & Heard, G., 2002. The yeast ecology of wine grapes. In: Ciani, M. (Ed.), Biodiversity and Biotechnology of Wine Yeasts. Research Signpost, Kerala, India, pp. 1–17.

Florenzano, G., Balloni, W. & Materassi R., 1977. Contributo alla ecologia dei lieviti *Schizosaccharomyces* sulle uve. Vitis 16, 38–44.

Fonseca, A., Inácio, J., 2006. Phylloplane yeasts. In: Péter, G., Rosa, C. (Eds.), Biodiversity and Ecophysiology of Yeasts. Springer, Berlin Heidelberg, pp. 263–301.

Francis, M.J.O. & Allcock, C., 1969. Geraniol □–D–glucoside: occurrence and synthesis in rose flowers. Phytochemistry 8, 1339–1347.

Garcia, A., Carcel, C., Dulau, L., Samson, A., Aguera, E., Agosin, E. & Gunata, Z., 2002. Influence of a mixed culture with *Debaryomyces vanriji* and *Saccharomyces cerevisiae* on the volatiles of a Muscat wine. J. Food Sci. 67, 1138–143.

Gil, J.V., Mateo, J.J, Jiménez, M., Pastor, A. & Huerta. T., 1996. Aroma compounds in wine as influences by apiculate yeasts. J. Food Sci 61, 1247–1249.

Giudici, P. & Pulvirenti, A., 2002. Molecular methods for identification of wine yeasts. In: Ciani, M. (Ed.), Biodiversity and Biotechnology of Wine Yeasts. Research Signpost, Kerala, India, pp. 35–52.

Gonde, P., Ratomahenina, R., Arnaud, A. & Galzy, P., 1985. Purification and properties of the exocellular □–glucosidase of *Candida wickerhamii* (Zikes) Meyer and Yarrow capable of hydrolysing soluble cellodextrines. Can. J. Biochem. Cell Biol. 63, 1160–1166.

Gonzalez, R., Quiros, M. & Morales, P., 2013. Yeast respiration of sugars by non–*Saccharomyces* yeast species: a promising and barely explored approach to lowering alcohol content of wines. Trends Food Sci. Technol. 29, 55–61.

Grimaldi, A., Bartowsky, E. & Jiranek,V., 2005. A survey of glycosidase activities of commercial wine strains of *Oenococcus oeni*. Int. J. Food Microbiol. 105, 233–234.

Grossmann, M., Rapp, A. & Rieth, W., 1987. Enzymatische freisetzung gebundener aromastoffe in wein. Dtsch. Lebensm. Rdsch. 83, 7–12.

Gueguen, Y., Chemardin, P., Labrot, P., Arnaud, A. & Galzy, P., 1997a. Purification and characterization of an intracellular β–glucosidase from a new strain of *Leuconostoc mesenteroides* isolated from cassava. J. Appl. Microbiol. 82, 469–476.

Gueguen, Y., Chemardin, P., Pien, S., Arnaud, A. & Galzy, P., 1997b. Enhancement of aromatic quality of Muscat wine by the use of immobilized □–glucosidase. J. Biotechnol. 55, 151–156.

Guillamón J.M., Sabaté J., Barrio E., Cano J. & Querol, A., 1998. Rapid identification of wine yeasts species based on RFLP analysis of the ribosomal ITS regions. Arch. Microbiol. 169, 387–392.

Gunata, Z., Bitteur, S., Brillouet, J.–M., Bayonove, C. & Cordonnier, R., 1988. Sequential enzymic hydrolysis of potentially aromatic glycosides from grape. Carbohydr. Res. 184, 139–149.

Gunata, Z., Bayonove, C., Tapiero, C. & Cordonnier, R., 1990a. Hydrolysis of grape monoterpenyl □–D–glucosides by various □–glucosidases. J. Agric. Food Chem. 38, 1232–1236.

Gunata, Z., Brillouet, J.M., Voirin, S., Baumes, B. & Cordonnier, R., 1990b. Purification and some properties of an □–L–arabinofuranosidase from *Aspergillus niger*. Action on grape monoterpenyl arabinofuranosyl glucosidases. J. Agric. Food Chem. 38, 772–776.

Gunata, Z., Dugelay, I., Sapis, J.C., Baumes, R. & Bayonove, C., 1990c. Action des glycosidases exogènes au cours de la vinification: liberation de l'arôme à partir des précurseurs glycosidiques. J. Int. Sci. Vigne Vin 24, 133–144.

Gunata Z., 1993. Etude et exploitation par voie enzymatize des precurseurs d'arômes du raisin de nature glycosidique. Revue des Oenologues 74, 22–27.

Gunata, Z., Dugelay, I., Sapis, J.C., Baumes, R. & Bayonove, C., 1993. Role of the enzymes in the use of the flavour potential from grape glycosides in winemaking. In: Schreier, P. & Winterhalter, P. (eds) Progress in Flavour Precursor Studies. Carol Stream IL, Allured Publ. pp 219–234.

Gunata, Z., Blondeel, C., Vallier, M.J., Lepoutre, J.P., Sapis, J.C., Watanabe, N., 1998. An endoglycosidase from grape berry skin of *cv* M. Alexandrie, hydrolysing potentially aromatic dissacharide glycosides. J. Agric. Food Chem. 46, 2748–2753.

Gunata, Z.. 2002. Flavor enhancement in fruit juices and derived beverages by exogenous glycosidases and consequences of the use of enzyme preparations. In: Whitaker, J.R. (ed) Handbook of Food Enzymology. Marcel Dekker, New York pp 303–330

Ha, M., Bekhit Ael, D., Carne, A. & Hopkins, D.L., 2013. Comparison of the proteolytic activities of new commercially available bacterial and fungal proteases toward meat proteins. J. Food Sci. 78, 170–7.

Heard, G.M. & Fleet, G.H., 1986. Evaluation of selective media for enumeration of yeasts during wine fermentation. J. Appl. Bacteriol. 60, 477–481.

Henick–Kling, T., Edinger, W., Daniel, P. & Monk, P., 1998. Selective effects of sulfur dioxide and yeast starter culture addition on indigenous yeast populations and sensory characteristics of wine. J. Appl. Microbiol. 84, 865–876.

Henschke, P.A. & Jiranek, V., 1993. Hydrogen sulfide formation during fermentation: effect of nitrogen composition in model grape must. In: *International symposium on nitrogen in grapes and wine*, pp 172–184. American Society for Enology and Viticulture Seattle, Davis.

Hernández, A., Martín, A., Aranda, E., Pérez Nevado, F. & Córdoba, M.G., 2007. Identification and characterization of yeasts isolated from the elaboration of seasoned green table olives. Food Microbiol. 24, 346–351.

Hernandez, L.F., Espinosa, J.C., Fernandez, M. & Briones, A., 2003. □–Glucosidase activity in a *Saccharomyces cerevisiae* wine strain. Int. J. Food Microbiol. 80, 171–176.

Hock, R., Benda, J. & Schreier, P., 1984. Formation of terpenes by yeasts during alcoholic fermentation. Z. Lebensm. Unters. Forsch. 179, 450–452.

Hrčková, M., Rusňáková, M. & Zemanovič, J., 2002. Enzymatic hydrolysis of defatted soy flour by three different proteases and their effect on the functional properties of resulting protein hydrolysates. Czech J. Food Sci. 20, 7–14.

Jackson, R.S., 2000. Wine Science – Principles, Pratice, Perception. Academic Press, San Diego, CA.

James, S.A., Cai, J., Roberts, I.N. & Collins, M.D., 1997. A phylogenetic analysis of the genus saccharomyces based on 18S rRNA gene sequences: description of *Saccharomyces kunashirensis* sp. nov. and *Saccharomyces martiniae* sp. nov. Int. J. Syst. Bacteriol. 47, 453–460.

Jindamorakot, S., Ninomiya, S., Limtong, S., Yongmanitchai, W., Tuntirungkij, M., Potacharoen, W., Tanaka, K., Kawasaki, H. & Nakase, T., 2009. Three new species of bipolar budding yeasts of the genus *Hanseniaspora* and its anamorph *Kloeckera* isolated in Thailand. FEMS Yeast Res. 9, 1327–1337.

Jolly, N., 2006. Die voorkoms van apikulate giste in druiween mosmonsters in die Robertson area. Wynboer Tegnies Mei, 68–70.

Jolly, N.P., Varela, C., Pretorius, I.S., 2014. Not your ordinary yeast: Non–*Saccharomyces* yeasts in wine production uncovered. FEMS Yeast Res. 14, 215–237.

Kish, S., Sharf, R. & Margalith, P., 1983. A note on a selective medium for wine yeasts. J. Appl. Microbiol. 55, 177–179.

Kurtzman C.P., O'Donell K. & Smith, M.T., 1994. Phylogeny of the yeast genera *Hanseniaspora* (anamorph *Kloeckera*), *Dekkera* (anamorph *Brettanomyces*), and *Eeniella* as inferred from partial 26s ribosomal DNA nucleotide sequences. Int. J. Syst. Bacteriol. 44, 781–786.

Kurtzman, C.P. & Robnett, C.J., 1998. Identification and phylogeny of ascomycetous yeast from analysis of nuclear large subunit (26S) ribosomal DNA partial sequences. Antonie van Leeuwenhoek 73, 331–371.

Kurtzman, C., Fell, J.W. & Boekhout, T., 2011. The yeasts: a taxonomic study, 5th edn. Elsevier. London, United Kingdon.

Kutyna, D.R., Varela, C., Henschke, P.A., Chambers, P.J. & Stanley, G.A., 2010. Microbiological approaches to lowering ethanol concentration in wine. Trends Food Sci. Technol 21, 293–302.

Lachance, M.A., 2003. The Phaff school of yeast ecology. Int. Microbiol. 6, 163–167.

Lagace, L.S. & Bisson, L.F., 1990. Survey of yeast acid proteases for effectiveness of wine haze reduction. Am. J. Enol. Vitic. 41, 147–155.

Lambrechts, M.G. & Pretorius, I.S., 2000. Yeast and its importance to wine aroma. S Afr. J. Enol. Vitic. 21, 97–129.

Leclerc, M., Arnaud, A., Ratomahenina, R., Galzy, P. & Nicolas, M., 1984. The enzyme system in a strain of *Candida wickerhamii* Meyer and Yarrow participating in the hydrolysis of cellodextrins. J. Gen. Appl. Microbiol. 30, 509–521.

López de Lerma, N. & Peinado, R.A. 2011. Use of two osmoethanol tolerant yeast strain to ferment must from Tempranillo dried grapes. Effect on wine composition. Int. J. Food Microbiol. 145, 342–348.

Loureiro, V., Malfeito–Ferreira, M. & Carreira, A., 2004. Detecting spoilage yeasts. In: Steele, R. (Ed.), Understanding and Measuring the Shelf–life of Food. Woodhead Publishers, Cambridge, pp. 233–288.

Madrigal, T., Maicas, S. & Mateo, J.J., 2013. Glucose and ethanol tolerant enzymes produced by *Pichia* (*Wickerhamomyces*) isolates from enological ecosystems. Am. J. Enol. Vitic. 64, 126–133.

Maicas, S. & Mateo, J.J. 2005. Hydrolysis of terpenyl glycosides in grape juice and other fruit juices: a review. Appl. Microbiol. Biotechnol. 67, 322–355.

Manitto, P., 1980. Byosynthesis of natural products, Ellis Horwood Publisher, Chichester.

Manzanares, P., Ramon, D. & Querol, A., 1999. Screening of non–*Saccharomyces* wine yeasts for the production of β–D–xylosidase activity. Int. J. Food Microbiol. 46, 105–112.

Martini, A., Federici, F. & Rosini, G., 1980. A new approach to the study of yeast ecology of natural substrates. Can. J. Microbiol. 26, 856–859.

Mateo, J.J., Jiménez, M. Huerta, T. & Pastor, A., 1991. Contribution of different yeasts isolated from musts of Monastrell grapes to the aroma of wine. Int. J. Food Microbiol. 14, 153–160.

Mateo, J.J., Jiménez, M. Huerta, T & Pastor, A., 1992. Comparison of volatiles produced by four *Saccharomyces* strains isolated from Monastrell musts. Am. J. Enol. Vitic. 43, 206–209.

Mateo, J.J. & Di Stefano, R., 1997. Description of the β–glucosidase activity of wine yeasts. Food Microbiol. 14, 583–591.

Mateo, J.J. & Jimenez, M., 2000. Monoterpene in grape juice and wines. J. Chromatogr. A 881, 557–567.

Mateo, J.J., Peris, L., Ibañez, C. & Maicas, S., 2011. Characterization of glycolytic activities from non–*Saccharomyces* yeasts isolated from Bobal musts. J. Ind. Microbiol. Biotechnol. 38, 347–354.

Mendes Ferreira, A., Clımaco, M.C. & Mendes Faia, A., 2001. The role of non–*Saccharomyces* species in releasing glycosidic bound fraction of grape aroma components–a preliminary study. J. Appl. Microbiol. 91, 67–71.

Moreira, N., Mendes, F. Hogg, T. & Vasconcelos, I., 2005. Alcohols, esters and heavy sulphur compounds production by pure and mixed cultures of apiculate wine yeasts. Int. J. Food Microbiol. 103, 285–294.

Muyzer, G., de Waal, E.C. & Uitterlinden, A.G., 1993. Profiling of complex microbial populations by denaturing gradient gel electrophoresis analysis of polymerase chain reaction–amplified genes coding for 16S rRNA. Appl. Environ. Microbiol. 59, 695–700.

Nieuwoudt, H.H., Prior, B.A., Pretorius, I.S. & Bauer, F.F., 2002. Glycerol in South African table wines: an assessment of its relationship to wine quality. S. Afr. J. Enol. Vitic. 23: 22–30.

Ogawa, K., Yasuyuki, I., Guo, W., Watanabe, N., Usui, T., Dong, S., Tong, Q. & Sakata, K., 1997. Purification of a □–primeverosidase concerned with alcoholic aroma formation in tea leaves (*Cv. Shuixian*) to be processed to oolong tea. J. Agr.c. Food Chem. 45, 877–882.

Oliveira, J.M., Oliveira, P., Baumes, R.L. & Maia, M.O., 2008. Volatile and glycosidically bound composition of Loureiro and Alvarinho wines. Food Sci. Technol. Int. 14, 341–353.

Oliveira, R.Q., Rosa, C.A., Uetanabaro, A.P., Azeredo, A., Neto, A.G. & Assis, S.A., 2009. Polygalacturonase secreted by yeasts from Brazilian semi–arid environments. Int. J. Food Sci. Nutr. 60 Suppl.7, 72–80.

Ozcan, T. & Kurdal, E., 2012. The effects of using a starter culture, lipase, and protease enzymes on ripening of Mihalic cheese. Int. J. Dairy Sci.Technol. 65, 585–593.

Pando, R., Lastra, A. & Suarez, B., 2012. Screening of enzymatic activities in non–*Saccharomyces* cider yeast. J. Food Biochem. 36, 683–689.

Paraggio, M., 2004. Biodiversity of a natural population of *Saccharomyces cerevisiae* and *Hanseniaspora uvarum* from Aglianico del Vulture. Food Technol. Biotechnol. 42, 165–168.

Pocock, K.F. & Waters, E.J., 2006. Protein haze in bottled white wines: How well do stability tests and bentonite fining trials predict haze formation during storage and transport? Austr. J. Grape Wine Res. 12, 212–220.

Prakitchaiwattana, C.J., Fleet, G.H. & Heard, G.M., 2004. Application and evaluation of denaturing gradient gel electrophoresis to analyse the yeast ecology of wine grapes. FEMS Yeast Res. 4, 865–877.

Pretorius, I.S., Van der Westhuizen, T.J. & Augustyr, O.P.H., 1999. Yeast biodiversity in vineyards and wineries and its importance to the South African wine industry. S. Afr. J. Enol. Vitic. 20, 61–74.

Pretorius, I.S., 2003. The genetic analysis and tailoring of wine yeasts. Topics in current genetics, Vol. 2 (De Winde JH, ed.), pp. 99–141. Springer–Verlag, Berlin.

Prior, B.A., Toh, T.H., Jolly, N., Baccari, C.L. & Mortimer, R.K., 2000. Impact of yeast breeding for elevated glycerol production on fermentation activity and metabolite formation in Chardonnay. S. Afr. J. Enol. Vitic. 21, 92–99.

Rapp, A. & Mandery, H., 1986. Wine aroma. Experientia 42, 873–884.

Rawlings, N.D. & Bateman, A., 2009. Pepsin homologues in bacteria. BMC Genomics 10, 437–448.

Rawlings, N.D., Barrett, A.J. & Bateman, A., 2009. MEROPS: the peptidase database. Nucl. Acids Res. 38, Database issue D227–D233.

Rawlings, N.D., Barrett, A.J. & Bateman, A., 2011. Asparagine peptide lyases. A seventh catalytic type of proteolytic enzymes. J. Biol. Chem. 286, 38321–38328.

Razungles, A., Bayonove, C.L., Cordonnier, R.E. & Baumes, R.L., 1987. Investigation on the carotenoids of the mature grape Vitis 26, 183–191

Renouf, V. & Lonvaud–Funel, A., 2007. Development of an enrichment medium to detect *Dekkera/Brettanomyces bruxellensis*, a spoilage wine yeast, on the surface of grape berries. Microbiol. Res. 162, 154–167.

Renzettia, S. & Arendta, E.K., 2009, Effect of protease treatment on the baking quality of brown rice bread: From textural and rheological properties to biochemistry and microstructure. J. Cereal Sci. 50, 22–28.

Restuccia, C., Pulvirenti, A., Caggia, C. & Giudici, P., 2002. A β–glucosidase positive strain of *Saccharomyces cerevisiae* isolated from grape must. Ann. Microbiol. 52, 47–53.

Ribereau–Gayon, P., Glories, Y., Maujean, A. & Dubourdieu, D., 2000. Handbook of Enology, Vol. 2. John Wiley & Sons Ltd, Chichester, pp. 187–206.

Riccio, P., Rossano, R., Vinella, M., Domizio, P., Zito, F., Sanseverino, F., D'Elia, A. & Rosi, I., 1999. Extraction and immobilization in one step of two □–glucosidases released from a yeast strain of *Debaryomyces hansenii*. Enz. Microb. Technol. 24, 123–129.

Rodrigues, F., Goncalves, G., Pereira–da–Silva, S., Malfeito–Ferreira, M. & Loureiro, V., 2001. Development and use of a new medium to detect yeasts of the genera *Dekkera/Brettanomyces*. J. Appl. Microbiol. 90, 588–599.

Rodríguez, F., Arroyo, F.N., Lopez, A., Bautista, J. & Garrido, A., 2010. Lipolytic activity of the yeast species associated with the fermentation/storage phase of ripe olive processing. Food Microbiol. 27, 604–612.

Rodríguez, M.E., Lopes, C.A., Barbagelata, R.J., Barda, N.B. & Caballero, A.C., 2010. Influence of *Candida pulcherrima* Patagonian strain on alcoholic fermentation behaviour and wine aroma. Int. J. Food Microbiol. 138, 19–25.

Rojas, V. Gil, J.V. Pinaga, F. & Manzanares P., 2001. Studies on acetate ester production by non–*Saccharomyces* wine yeasts. Int. J. Food Microbiol. 70, 283–289.

Romano, P. & Suzzi, G., 1996. Origin and production of acetoin during wine yeast fermentation (mini review). Appl. Environ. Microbiol. 62, 309–315.

Romano, P., Fiore, C., Paraggio, M., Caruso, M. & Capece, A. 2003. Function of yeast species and strains in wine flavor. Int. J. Food Microbiol. 86, 169–180.

Romero, A.M., Mateo, J.J. & Maicas, S., 2012. Characterization of an ethanol tolerant 1,4–β–xylosidase produced by *Pichia membranifaciens*. Lett. Appl. Microbiol. 55, 354–361.

Rosi, I. & Costamagna, L., 1987. Screening for extracellular acid protease(s) production by wine yeasts. J. Inst. Brewing 93, 322–324.

Rosi, I., Vinella, M. & Domizio, P., 1994. Characterization of β–glucosidase activity in yeasts of oenological origin. J. Appl. Bacteriol. 77, 519–527.

Sadoudi, M., Tourdot–Marechal, R., Rousseaux, S., Steyer, D., Gallardo–Chacon, J.J., Ballester, J., Vichi, S., Guerin–Schneider, R., Caixach, J. & Alexandre, H., 2012. Yeast–yeast interactions revealed by aromatic profile analysis of Sauvignon Blanc wine fermented by single or co–culture of non–*Saccharomyces* and *Saccharomyces* yeasts. Food Microbiol. 32, 243–253.

Sampaio, J.P. & Gonçalves, P., 2008. Natural populations of *Saccharomyces kudriavzevii* in Portugal are associated with Oak bark and are sympatric with *S. cerevisiae* and *S. paradoxus*. Appl. Environ. Microbiol. 74, 2144–2152.

Scanes, K.T., Hohmann, S. & Prior, B.A., 1998. Glycerol production by the yeast *Saccharomyces cerevisiae* and its relevance to wine: a review. S. Afr. J. Enol. Vitic. 19, 15–21.

Schuller, D., Côrte–Real, M. & Leão, C., 2000. A differential medium for the enumeration of the spoilage yeast *Zygosaccharomyces bailii* in wine. J. Food Protect. 63, 1570–1575.

Sefton, M.A & Williams, P.J., 1991. Generation of oxidation artifacts during the hydrolysis of norisoprenoid glycosides by fungal enzyme preparations. J. Agric. Food Chem. 39, 1994–1997.

Spagna, G., Barbagallo, R.N., Palmeri, R., Restuccia, C. & Giudici, P., 2002. Properties of endogenous β–glucosidase of a *Pichia anomala* strain isolated from Sicilian musts and wines. Enzyme Microb. Technol. 31, 1036–1041.

Strauss, M.C.A., Jolly, N.P., Lambrechts, M.G. & van Rensburg, P., 2001. Screening for the production of extracellular hydrolytic enzymes by non–*Saccharomyces* wine yeasts. J. Appl. Microbiol. 91, 182–190.

Sturm, J., Grossmann, M. & Schnell, S., 2006. Influence of grape treatment on the wine yeast populations isolated from spontaneous fermentations. J. Appl. Microbiol. 101, 1241–1248.

Swangkeaw J., Vichitphan, S. Butzke, C.E. & Vichitphan, K., 2011. Characterization of β–glucosidases from *Hanseniaspora* sp. and *Pichia anomala* with potentially aroma–enhancing capabilities in juice and wine. World J. Microbiol. Biotechnol. 27, 423–430.

Swiegers, J.H. & Pretorius, I.S., 2005. Yeast modulation of wine flavour. Adv Appl Microbiol 57, 131–175.

Swiegers, J.H., Bartowsky, E.J., Henschke, P.A. & Pretorius, I.S., 2005. Yeast and bacterial modulation of wine aroma and flavour. Austr. J. Grape Wine Res. 11, 139–173.

Swiegers, J.H., Kievit, R.L., Siebert, T., Lattey, K.A., Bramley, B.R., Francis, I.L., King, E.S. & Pretorius, I.S., 2009. The influence of yeast on the aroma of Sauvignon blanc wine. Food Microbiol. 26, 204–211.

Tremacoldi, C.R. & Carmona, E.C., 2005. Production of extracellular alkaline proteases by *Aspergillus clavatus*. World J. Microbiol. Biotechnol. 21, 169–172.

Tyndall, J.D.A., Nall, T. & Fairlie, D.P., 2005. Proteases universally recognize β–strands in their active sites. Chemical Rev. 105, 973–1000.

Ubeda–Iranzo, J.F., Briones–Perez, A.I. & Izquierdo–Canas, P.M., 1998. Study of the oenological characteristics and enzymatic activities of wine yeasts. Food Microbiol. 15, 399–406.

Ubeda, J. & Briones, A., 2000. Characterization of differences in the formation of volatiles during fermentation within synthetic and grape musts by wild *Saccharomyces* strains. LWT – Food Sci. Technol. 33, 408–414.

Ugliano, M., 2009. Enzymes in winemaking. In: Wine Chemistry and Biochemistry. Moreno–Arribas, M.V., Polo, C. (Eds). Springer Science–Business Media, Adelaide, Australia. pp 103–126.

Urso, R., Rantsiou, K., Dolci, P., Rolle, L., Comi, G. & Cocolin, L., 2008. Yeast biodiversity and dynamics during sweet wine production as determined by molecular methods. FEMS Yeast Res. 8, 1053–1062.

Viana, F., Gil, J.V., Vallés, S. & Manzanares, P., 2009. Increasing the levels of 2–phenylethyl acetate in wine through the use of a mixed culture of *Hanseniaspora osmophila* and *Saccharomyces cerevisiae*. Int. J. Food. Microbiol. 135, 6–75.

Vidal, S., Francis, L., Williams, P., Kwiatkowski, M., Gawel, R., Cheynier, W. & Waters, E., 2004. The mouth–feel properties of polysaccharides and anthocyanins in a wine like medium. Food Chem. 85, 519–525.

Waters, E.J., Alexander, G., Muhlack, R., Pocock, K.F., Colby, C., O'Neill, B.K., Høj, P.B. & Jones, P., 2005. Preventing protein haze in bottled white wine. Austr. J. Grape Wine Res. 11, 215–225.

White, T.J., Bruns, T., Lee, S. & Taylor, J., 1990. Amplification and direct sequencing of fungal ribosomal RNA genes for phylogenetics. In: Innis, M.A., Gelfand, D.H., Sninsky, J.J. & White, T.J. (Eds) PCR Protocols. A Guide to Methods and Applications, Academic Press, San Diego, CA, USA, pp. 315–322.

Williams, P.J., Strauss, C.R., Wilson, B. & Massy–Westropp, R.A., 1982. Use of C_{18} reversed–phase liquid chromatography for the isolation of monoterpene glycosides and nor–isoprenoid precursors from grape juice and wines. J. Chromatogr. 235, 471–480.

Wilson, B., Strauss, C.R. & Williams, P.J., 1986. The distribution of free and glycosidically bound monoterpenes among skin, juice and pulp fractions of some white grape varieties. Am. J. Enol. Vitic. 37, 107–111.

Yanagida, F., Ichinose, F., Shinohara, T. & Goto, S., 1992. Distribution of wild yeasts in the white grape varieties at Central Japan. J. Gen. Appl. Microbiol. 38, 505–509.

Yanai, T. & Sato, M., 1999. Isolation and properties of β–glucosidase produced by *Debaryomyces hansenii* and its application in winemaking. Am. J. Enol. Vitic. 50, 231–235.

Zironi, R., Romano, P., Suzzi, G., Battistutta, F. & Comi, G., 1993. Volatile metabolites produced in wine by mixed and sequential cultures of *Hanseniaspora guilliermondii* or *Kloeckera apiculata* and *Saccharomyces cerevisiae*. Biotechnol. Lett. 15, 235–238.

Zott, K., Thibon, C., Bely, M., Lonvaud–Funel, A., Dubourdieu, D. & Masneuf–Pomarede, I., 2011. The grape must non–*Saccharomyces* microbial community: impact on volatile thiol release. Int. J. Food Microbiol. 151, 210–215.

Printed by Books on Demand GmbH, Norderstedt / Germany